Lecture Notes in Mathematics 1589

Editors:
A. Dold, Heidelberg
B. Eckmann, Zürich
F. Takens, Groningen

Subseries: Fondazione C.I.M.E., Firenze

Advisor: Roberto Conti

J. Bellissard   M. Degli Esposti   G. Forni
S. Graffi   S. Isola   J. N. Mather

Transition to Chaos in Classical and Quantum Mechanics

Lectures given at the 3rd Session of the
Centro Internazionale Matematico Estivo
(C.I.M.E.) held in Montecatini Terme, Italy,
July 6-13, 1991

Editor: S. Graffi

Fondazione
C.I.M.E.

Springer-Verlag
Berlin Heidelberg New York
London Paris Tokyo
Hong Kong Barcelona
Budapest

Authors

Jean Bellissard
Laboratoire de Physique Quantique
Université Paul Sabatier
118, route de Narbonne
F-31062 Toulouse Cedex, France

Mirko Degli Esposti
Giovanni Forni
Sandro Graffi
Stefano Isola
Dipartimento di Matematica
Università degli di Studi di Bologna
Piazza di Porta S. Donato, 5
I-40127 Bologna, Italy

John N. Mather
Department of Mathematics
Princeton University
Princeton, NJ 08544, USA

Editor

Sandro Graffi
Dipartimento di Matematica
Università degli Study di Bologna
Piazza di Porta S. Donato, 5
I-40127 Bologna, Italy

Mathematics Subject Classification (1991): 58F, 58F05, 58F15, 58F36, 81Q, 81Q05, 81Q20, 81Q50, 81S, 81S05, 81S10, 81S30

ISBN 3-540-58416-1 Springer-Verlag Berlin Heidelberg New York

CIP-Data applied for

Typesetting: Camera ready by author
SPIN: 10130140   46/3140-543210 - Printed on acid-free paper

FOREWORD

This volume collects the texts of two series of 8 lectures, and the expanded version of a seminar, given at the ⟨C.I.M.E. Session on "Transition to Chaos in Classical and Quantum Systems", which took place at the Villa "La Querceta" in Montecatini, Italy, from July 6 to July 13, 1991.

The purpose of the Session was to give a broad survey of the mathematical problems and techniques, as well as of some of the most relevant physical motivations, which arise in the study of the stochastic behaviour, if any, of deterministic dynamical systems both in classical and quantum mechanics.

The transition to chaos in the most relevant and widely studied examples of classical dynamical systems, the area preserving maps, is thoroughly covered in the first series of lectures, delivered by Professor John Mather and written in collaboration with Dr. Giovanni Forni. In particular the reader can find in this text an up-to-date version of the well known Aubry-Mather theory. The lectures of Professor Jean Bellissard cover in turn, in addition to his algebraic approach to the classical limit, the behaviour of the quantum counterpart of the above systems, with particular emphasis on localization, and on qualitative as well as quantitative properties of the spectra of the relevant Schrödinger operators in classically chaotic regions. They can be therefore considered an exhaustive introduction to the mathematical aspects of the so-called "quantum chaos". The third series of lectures, delivered by Professor Anatole Katok, covered the basic stochastic properties of classical dynamical systems and some of their most recent developments. Unfortunately Professor Katok could not find the time to write up the text of his course.

A very prominent role in describing the chaotic behaviour of classical dynamical systems is played, as discussed also in Professor's Katok lectures, by the proliferation and equidistribution of the unstable periodic orbits of increasing period. An overview of recent results in this direction, and of their intimate connection to the problem of the classical limit of the quantized toral symplectomorphisms, is contained in an outgrowth of a seminar held by M.Degli Esposti, written in collaboration with S.Isola and the Editor.

Bologna, April 1994

Sandro Graffi

TABLE OF CONTENTS

Non Commutative Methods in Semiclassical Analysis

Jean Bellissard

Laboratoire de Physique Quantique Université Paul Sabatier
118, route de Narbonne F-31062 Toulouse Cedex, France

Contents

1   The kicked rotor problem

One considers a spinning particle submitted to rotate around a fixed axis. Let $\theta \in \mathbf{T} = \mathbf{R}/2\pi\mathbf{Z}$ be its angle of rotation, $L \in \mathbf{R}$ its angular momentum, I its moment of inertia, μ its magnetic moment, and B a uniform magnetic field parallel to the axis of rotation. Its kinetic energy is given by :

$$\mathcal{H}_0 = \frac{L^2}{2I} + \mu B L , \tag{1}$$

We assume that this system is kicked periodically in time according to the following Hamiltonian :

$$\mathcal{H} = \frac{L^2}{2I} + \mu B L + k \cos(\theta) \sum_{n \in \mathbf{Z}} \delta(t - nT) . \tag{2}$$

where T is the period of the kicks, and k is a coupling constant representing the kicks strength. Here δ is the Dirac measure. Classically the motion is provided by the solution of the Hamilton-Jacobi equations :

$$\frac{d\theta}{dt} = \frac{\partial \mathcal{H}}{\partial L} \qquad \frac{dL}{dt} = -\frac{\partial \mathcal{H}}{\partial \theta} . \tag{3}$$

Between two kicks, $\partial \mathcal{H}/\partial \theta = 0$, so that L is constant whereas θ varies linearly in time. When the kick is applied, L changes suddenly according to $L(nT + 0) = L(nT - 0) + k \sin(\theta)$. If we set :

$$A_n = T \left(\frac{L(nT - 0)}{I} + \mu B \right) \qquad \theta_n = \theta(nT - 0) , \tag{4}$$

the equation of motion can be expressed as :

$$A_{n+1} = A_n + K \sin(\theta_n) \qquad \theta_{n+1} = \theta_n + A_{n+1} \bmod 2\pi , \tag{5}$$

where K is the dimensionless coupling strength namely :

$$K = \frac{kT}{I} . \tag{6}$$

The phase space is the cylinder $\mathcal{C} = \mathbf{T} \times \mathbf{R}$, if A is considered as a real number. If we set

$$f(\theta, A) = (\theta', A') \qquad \theta' = \theta + A + K \sin(\theta) \qquad A' = A + K \sin(\theta) , \tag{7}$$

the solution of the equation of motion can be written as :

$$(\theta_{n+1}, A_{n+1}) = f(\theta_n, A_n) . \tag{8}$$

f is an analytic diffeomorphism of the cylinder $\mathcal{C}$, which is area preserving, namely $d\theta' \wedge dA' = d\theta \wedge dA$, and a twist map, namely $\partial \theta'/\partial A > 0$, which preserves the ends (see the course of John Mather in this issue). We remark that f also commutes with the translation $A \mapsto A + 2\pi$ of the action variable A in such a way that it also defines a map of the 2-torus $\mathbf{T}^2$.

The orthodox way of quantizing this model consists in choosing the Hilbert space $\mathcal{K} = L^2(\mathbf{T}, d\theta/2\pi)$ as the state space, and replacing L and θ by operators as follows :

$$\mathbf{L} = \frac{\hbar}{i} \frac{\partial}{\partial \theta} \qquad \mathcal{V} = \text{multiplication by } \mathcal{V}(\theta) , \qquad (9)$$

whenever $\mathcal{V}$ is a continuous 2π -periodic function of the variable θ . Quantum Mechanics requires using a new parameter $\hbar$, the Planck constant which gives rise to a new dimensionless parameter :

$$\gamma = \frac{\hbar T}{I} = 4\pi \frac{\nu_{\mathrm{QM}}}{\nu_{\mathrm{CL}}} , \qquad (10)$$

where $\nu_{\mathrm{CL}} = 1/T$ is the kicks frequency, whereas ν_{QM} is the eigenfrequency of the free quantum rotor in a zero magnetic field. To compute the motion, we need to solve Schrödinger's equation, namely, we look for a path $t \in \mathbf{R} \mapsto \psi_t \in \mathcal{K}$ such that :

$$i\hbar\psi_t = H(t)\psi_t \qquad H(t) = \frac{\mathbf{L}^2}{2I} + \mu B \mathbf{L} + k \cos(\theta) \sum_{n \in \mathbf{Z}} \delta(t - nT) . \qquad (11)$$

The δ-kicks may create a technical difficulty. To overcome it let us consider a smooth approximation δ_ϵ of δ given by a non negative L^1-function on $\mathbf{R}$ supported by $[0, \epsilon]$, with integral equal to 1. The solution can be given in term of a convergent Dyson expansion. Then letting ϵ converge to zero, we get the following result (see Appendix 1) :

Theorem 1 *The solution of (11) is given by the following evolution equation :*

$$\psi_{T-0} = F^{-1}\psi_{0-} \qquad F^{-1} = e^{-iA^2/2\gamma}e^{-iK\cos\theta/\gamma}e^{i\hat{y}} \qquad (12)$$

where

$$A = T\left(\frac{L}{I} + \mu B\right) , \qquad \hat{y} = (\mu B)^2 \frac{TI}{\hbar} . \qquad (13)$$

Let us also introduce the dimensionless magnetic field x :

$$x = -\mu BT \qquad \Rightarrow \qquad \hat{y} = \frac{x^2}{2\gamma} . \qquad (14)$$

The operators of the form $\mathcal{V}$ whenever $\mathcal{V}(\theta)$ is a continuous 2π -periodic function of the variable θ, can be obtained as the norm limit of polynomials in the operator

$$U = e^{i\theta} . \qquad (15)$$

In much the same way, one can quantize the action in the torus geometry by considering the operator :

$$V = e^{-iA} . \qquad (16)$$

U and V are two unitary operators satisfying the following commutation rule :

$$UV = e^{i\gamma}VU . \qquad (17)$$

The C^*-algebra generated by these two operators is the non commutative analog of the space of continuous functions on the 2-torus. By analogy with the commutative case, this algebra will be seen as the space of continuous functions on a virtual space, the "quantal phase space". Any such function will be the norm limit of polynomials of the form :

$$a = \sum_{\mathbf{m}\in\mathbf{Z}^2, |\mathbf{m}|\leq N} a(\mathbf{m})U^{m_1}V^{m_2}e^{-i\gamma m_1 m_2/2} , \tag{18}$$

where the $a(\mathbf{m})$'s are complex numbers. We denote by $\mathcal{A}_\gamma$ the norm closure of this algebra. Whenever $\gamma = 0$, this algebra coincides with the space $\mathcal{C}(\mathbf{T}^2)$ of continuous functions on the 2-torus. One remarks that $\cos(\theta) \in \mathcal{A}_\gamma$, but there is no way of writing $F_0 = \exp(iA^2/\gamma)$ as an element of $\mathcal{A}_\gamma$ since it is not periodic with respect to A. Therefore $F_0 \notin \mathcal{A}_\gamma$ in general. However, the following properties hold :

$$(\mathrm{i})F_0VF_0^{-1} = V \qquad (\mathrm{ii})F_0UF_0^{-1} = UV^{-1}e^{i\gamma/2} , \tag{19}$$

so that, setting $\beta_0(a) = F_0aF_0^{-1}$ for $a \in \mathcal{A}_\gamma$, β_0 defines an automorphism of $\mathcal{A}_\gamma$, which coincides for $\gamma = 0$ with the free rotation f_0 in $\mathbf{T}^2$, namely :

$$f_0(\theta, A) = (\theta + A, A) , \tag{20}$$

In particular if $\gamma \neq 0$, $a \in \mathcal{A}_\gamma$, we get :

$$\beta(a) = FaF^{-1} = e^{iK\cos(\theta)/\gamma}\beta_0(a)e^{-iK\cos(\theta)/\gamma} \in \mathcal{A}_\gamma , \tag{21}$$

which means that β is an automorphism of $\mathcal{A}_\gamma$.

At last, β admits a classical limit as $\gamma \mapsto 0$, namely the automorphism of $\mathcal{C}(\mathbf{T}^2)$ corresponding to the standard map (see section 3 below). For if $\mathcal{V} = K\cos(\theta)$, let us denote by $\mathcal{L}_v$ the "Liouville operator" defined by :

$$\mathcal{L}_v(a) = \frac{\mathcal{V}a - a\mathcal{V}}{i\gamma} , \tag{22}$$

the limit of $\mathcal{L}_v(a)$ as $\gamma \mapsto 0$ coincides with the Poisson bracket of $\mathcal{V}$ with a, and β can be written as :

$$\beta = e^{-\mathcal{L}_v} \circ \beta_0 . \tag{23}$$

To summarize, we have obtained an algebraic framework describing the quantal observables which is completely analogous to the classical description of the system, and which converges to the classical analog as $\gamma \mapsto 0$. In this framework,

(i) the observable algebra $\mathcal{A}_\gamma$ is the non commutative analog of the space $\mathcal{C}(\mathbf{T}^2)$ of continuous functions on the classical phase space $\mathbf{T}^2$.

(ii) the quantal evolution is described through the automorphism β of $\mathcal{A}_\gamma$ which admits the standard map as a classical limit.

Before leaving this section, let us describe the complementary point of view, given in wave Mechanics by the Feynman path integral, which happens to be exact and finite dimensional in this case.

Lemma 1 *If $\psi \in \mathcal{C}^\infty(\mathbf{T})$, then the following formula holds :*

$$\left(e^{-iA^2/2\gamma}\psi\right)(u) = e^{-i\pi/4}\int_{-\infty}^{+\infty}\frac{du'}{\sqrt{2\pi\gamma}}e^{i(u'-u-x)^2/2\gamma}e^{-ix^2/2\gamma}\psi(u') . \tag{24}$$

Proof : From (9)&(13), we get $A = -i\gamma\partial/\partial\theta - x$. If $\psi \in \mathcal{C}^\infty(\mathbf{T})$, let $(\psi_n)_{n\in\mathbf{Z}}$ its Fourier series, so that :

$$\left(e^{-iA^2/2\gamma}\psi\right)(\theta) = \sum_{n\in\mathbf{Z}} e^{-i(\gamma n-x)^2/2\gamma}\psi_n e^{in\theta} = \sum_{n\in\mathbf{Z}} \int_{-\pi}^{+\pi} \frac{d\theta'}{2\pi} e^{in(\theta-\theta')-i(\gamma n-x)^2/2\gamma}\psi(\theta') \ .$$

To compute the distribution kernel coming into this sum, we use the Poisson summation formula :

$$\sum_{n\in\mathbf{Z}} e^{in(\theta-\theta'+x)-i\gamma n^2/2} = \frac{e^{-i\pi/4}}{\sqrt{2\pi\gamma}} 2\pi \sum_{l\in\mathbf{Z}} e^{i(\theta-\theta'+x+2\pi l)^2/2\gamma} \ .$$

Now we perform the change of variables $u' = \theta' + 2\pi l$, $u = \theta$, and the sum over $l \in \mathbf{Z}$ will give rise to an integral over $\mathbf{R}$ with respect to u', leading to (24).

Using (12)&(24), we immediately get the following Feynman path integral representation :

Corollary 1 *For any $t \in \mathbf{N}$ and $\psi \in \mathcal{C}^\infty(\mathbf{T})$, we get :*

$$\left(F^{-t}\psi\right)(u) = \int_{\mathbf{R}^t} \frac{du_1 \cdots du_t}{(2\pi\gamma)^{t/2}} e^{-it\pi/4} e^{i(\sum_{s=1}^t (u_s-u_{s-1}-x)^2/2 - K\cos(u_s))/\gamma}\psi(u_t) \ , \qquad (25)$$

where $u_0 = u$, and the right-hand side defines a convergent oscillatory integral which is periodic of period 2π with respect to u.

Remark : The expression contained in the phase factor

$$S(u_1, \cdots, u_t; u_0, x) = \sum_{1\le s\le t-2} \left(\frac{(u_s - u_{s-1} - x)^2}{2} - K\cos(u_s) \right) \ , \qquad (26)$$

is nothing but the "Percival" Lagrangean or the "Frenkel-Kontorova" energy functional used by Aubry and Mather to describe the trajectories of the standard map. For indeed the stationnary points of such a Lagrangean are finite sequences $(u_s)_{1\le s\le t}$ satisfying the recursion relation :

$$2u_s - u_{s+1} - u_{s-1} + K\sin(u_s) = 0 \ , \ (1 \le s \le t-1) \ , \ u_t - u_{t-1} - x + K\sin(u_t) = 0 \ .$$

In particular if we set $p_s = u_s - u_{s-1}$ (for $1 \le s \le t$) we get $u_{s+1} = u_s + p_{s+1}$ for $0 \le s \le t-1$, and $p_{s+1} = p_s + K\sin(u_s)$ for $1 \le s \le t-1$, $x = p_t + K\sin(u_t)$, namely we recover the standard map (5) in $\mathbf{R}^2$ now instead of $\mathbf{T}^2$, for a trajectory $(\theta_0, A_0), \cdots, (\theta_t, A_t)$ such that $\theta_0 = u_0 \bmod 2\pi$, and $A_{t+1} = x \bmod 2\pi$.

2 The Rotation Algebra

2.1 The Polynomial Algebra $\mathcal{P}_I$

In this section we define properly the algebra $\mathcal{A}_\gamma$ and we will describe without proof its most important properties. We refer the reader to [BaBeFl] for more details. Actually given an interval I of $\mathbf{R}$, we will rather consider the algebra $\mathcal{A}_I$ which is roughly speaking the set of continuous sections of the continuous field $\gamma \in I \mapsto \mathcal{A}_\gamma$. The semiclassical limit will be included whenever I contains $\gamma = 0$.

Let I be a compact subset of $\mathbf{R}$. The polynomial algebra $\mathcal{P}_I$ is defined as follows :

- the elements of $\mathcal{P}_I$ are the sequences $(a(\mathbf{m}))_{\mathbf{m} \in \mathbf{Z}^2}$ with finite support, where for each $\mathbf{m} = (m_1, m_2) \in \mathbf{Z}^2$, $a(\mathbf{m}) : \gamma \in I \mapsto a(\mathbf{m}, \gamma) \in \mathbf{C}$ is a complex continuous function on I.

- $\mathcal{P}_I$ admits a natural structure of $\mathcal{C}(I)$-module by setting, for $a, b \in \mathcal{P}_I$, and $l \in \mathcal{C}(I)$:

$$(a+b)(\mathbf{m}) = a(\mathbf{m}) + b(\mathbf{m}) \qquad \lambda a(\mathbf{m}; \gamma) = \lambda(\gamma) a(\mathbf{m}; \gamma) . \tag{27}$$

- any element $a \in \mathcal{P}_I$ admits an adjoint a^* defined by :

$$a^*(\mathbf{m}; \gamma) = \overline{a(-\mathbf{m}; \gamma)} , \tag{28}$$

where $\overline{z}$ denotes the complex conjugate of z in $\mathbf{C}$.

-if $a, b \in \mathcal{P}_I$, their product is defined by :

$$(ab)(\mathbf{m}; \gamma) = \sum_{\mathbf{m}' \in \mathbf{Z}^2} a(\mathbf{m}'; \gamma) b(\mathbf{m} - \mathbf{m}'; \gamma) e^{i\gamma \mathbf{m}' \wedge (\mathbf{m} - \mathbf{m}')} , \tag{29}$$

where we have set if $\mathbf{m}', \mathbf{m}'' \in \mathbf{Z}^2$:

$$\mathbf{m}' \wedge \mathbf{m}'' = m_1' m_2'' - m_2' m_1'' . \tag{30}$$

- the topology on $\mathcal{P}_I$, is the direct sum topology obtained from the uniform norm on $\mathcal{C}(I)$.

Denoting by $\mathcal{P}_\gamma$ the algebra $\mathcal{P}_I$ whenever $I = \{\gamma\}$ it follows that $\mathcal{P}_\gamma = \mathcal{P}_{\gamma+4\pi}$. Moreover setting $\alpha(a) = ((-)^{m_1 m_2} a(m))_{\mathbf{m} \in \mathbf{Z}^2}$, α defines a $*$-isomorphism between $\mathcal{P}_\gamma$ and $\mathcal{P}_{\gamma+2\pi}$. Thus, as far as $\mathcal{P}_\gamma$ is concerned, one will consider that γ is defined mod. 2π. The same definition holds if we replace I by the torus $\mathbf{T}$ namely the continuous functions on I by the continuous 2π-periodic functions on $\mathbf{R}$. We will denote by $\mathcal{P}$ the corresponding algebra.

The following elements in $\mathcal{P}_I$ are remarkable :

$$\mathbf{I}(\mathbf{m}; \gamma) = \delta_{\mathbf{m},0} \qquad U(\mathbf{m}; \gamma) = \delta_{\mathbf{m},(1,0)} \qquad V(\mathbf{m}; \gamma) = \delta_{\mathbf{m},(0,1)} . \tag{31}$$

For indeed, $\mathbf{I}$ is the identity of $\mathcal{P}_I$ whereas U, V, are unitaries namely $UU^* = U^*U = VV^* = V^*V = \mathbf{I}$, and obey to the commutation rules (17). Moreover, $\mathcal{P}_I$ is algebraically generated by U, V as a $\mathcal{C}(I)$-algebra, namely if $a \in \mathcal{P}_I$, it can be written as :

$$a = \sum_{\mathbf{m}' \in \mathbf{Z}^2} a(\mathbf{m}) U^{m_1} V^{m_2} e^{-i\gamma m_1 m_2 / 2} .$$

It will be convenient to introduce the "Weyl operators" as follows :

$$W(\mathbf{m}) = U^{m_1} V^{m_2} e^{-i\gamma m_1 m_2 / 2} . \tag{32}$$

From the interpretation given in the previous section, it follows that $\mathcal{P}_I$ is the set of trigonometric polynomials over the "non-commutative" 2-torus. In particular if $I = \{0\}$, we recover the convolution algebra, which by Fourier transform is exactly the algebra of usual trigonometric polynomials.

The "evaluation" homomorphism η_γ is defined as the map from $\mathcal{P}_I$ into $\mathcal{P}_\gamma$ by :

$$\eta_\gamma(a) = (a(\mathbf{m}; \gamma))_{\mathbf{m} \in \mathbf{Z}^2} . \tag{33}$$

It is immediate to check that η_γ is a $*$-homomorphism, namely it is linear, and preserves the product and the adjoint.

2.2 Canonical calculus

Using the analogy with the space of trigonometric polynomials on the 2-torus, we now define some rules for the differential calculus.
The integral is given by the trace defined by :

$$\tau(a) = a(0) \in \mathcal{C}(I) . \tag{34}$$

We will denote by $\tau_\gamma(a)$ the value of $\tau(a)$ at γ. The trace τ is a linear module map from $\mathcal{P}_I$ into $\mathcal{C}(I)$ satisfying :

(i) positivity : $\tau(a^*a) = \sum_{m' \in \mathbf{Z}^2} |a(m)|^2 \geq 0$, $a \in \mathcal{P}_I$,
(ii) normalization : $\tau(\mathbf{I}) = 1$,
(iii) trace property : $\tau(ab) = \tau(ba)$, $a, b \in \mathcal{P}_I$.

We remark that the value of $\tau(a)$ at $\gamma = 0$ is the 0th Fourier coefficient of $\eta_0(a)$, namely the integral of its Fourier transform :

$$\tau(a)|_{\gamma=0} = \int_{\mathbf{T}^2} \frac{d\theta dA}{4\pi^2} a_{cl}(\theta, A) . \tag{35}$$

where a_{cl} is the Fourier transform of $\eta_0(a)$.
The angle average, is defined by the element $\langle a \rangle$ in $\mathcal{P}_I$ given by :

$$\langle a \rangle(\mathbf{m}) = \delta_{m_1,0} a(0, m_2) . \tag{36}$$

The map $a \mapsto \langle a \rangle$ is a module-map taking values in the commutative subalgebra $\mathcal{D}_I$ generated by V as a $\mathcal{C}(I)$-module. The usual Fourier transform permits to associate with any element b of $\mathcal{D}_I$ a continuous function of $(\gamma, A) \in I \times \mathbf{T}$ denoted by b_{av} as follows :

$$b_{av}(\gamma, A) = \sum_{m' \in \mathbf{Z}^2} b(0, m_2; \gamma) e^{-im_2 A} . \tag{37}$$

The mapping $b \in \mathcal{D}_I \mapsto b_{av} \in \mathcal{C}(I \times \mathbf{T})$, is a $*$-homomorphism, namely $(bc)_{av} = b_{av} c_{av}$ and $(b^*)_{av} = b^*_{av}$. We will say that $b \in \mathcal{D}_I$ is positive whenever b_{av} is positive. Using these definitions, the angle averaging satisfies :

(i)	positivity property :	$\langle a^*a \rangle \geq 0$, $a \in \mathcal{P}_I$
(ii)	projection property :	$\langle \langle a \rangle \rangle = \langle a \rangle$,
(iii)	normalization :	$\langle \mathbf{I} \rangle = 1$,
(iv)	conditional expectation :	$\langle ab \rangle = \langle a \rangle b$, $\langle ba \rangle = b \langle a \rangle$, if $b \in \mathcal{D}_I$, $a \in \mathcal{P}_I$.

$$\tag{38}$$

A differential structure is defined on $\mathcal{P}_I$ through the data of two $*$-derivations ∂_θ and ∂_A given by :

$$(\partial_\theta a)(\mathbf{m}) = im_1 a(\mathbf{m}) \qquad (\partial_A a)(\mathbf{m}) = im_2 a(\mathbf{m}) . \tag{39}$$

These two derivations ∂_μ (if $\mu = \theta, A$) actually commute and satisfy :

(i)	they are $\mathcal{C}(I)$−linear	
(ii)	$\partial_\mu(a^*) = (\partial_\mu a)^*$	$a \in \mathcal{P}_I$,
(iii)	$\partial_\mu(ab) = (\partial_\mu a) b + a (\partial_\mu b)$	$a, b \in \mathcal{P}_I$,
(iv)	$\partial_\theta U = iU$, $\partial_\theta V = 0$, $\partial_A U = 0$, $\partial_A V = -iV$.	

$$\tag{40}$$

Moreover one can exponentiate them, namely defining by $\{\rho_{\theta,A}; (\theta, A) \in \mathbf{T}^2\}$ as the 2-parameter group of $*$-automorphisms given by :

$$\rho_{\theta,A}(a)(\mathbf{m}) = e^{i(m_1\theta - m_2 A)}a(\mathbf{m}) , \qquad (41)$$

we get :

$$\partial_\mu a = \left(\frac{\partial \rho_{\theta,A}(a)}{\partial \mu}\right)_{\theta=A=0} \qquad \mu = \theta, A . \qquad (42)$$

Actually $\rho_{\theta,A}$ is a module-$*$-homomorphism such that $(\theta, A) \in \mathbf{T}^2 \mapsto \rho_{\theta,A}(a) \in \mathcal{P}_I$ is continuous and :

$$\rho_{\theta,A} \circ \rho_{\theta',A'} = \rho_{\theta+\theta',A+A'} , \qquad (43)$$

If $a, b \in \mathcal{P}_I$ their Poisson (or Moyal [Bou]) bracket $\{a, b\}$ is defined as follows :

$$\{a, b\}(\mathbf{m}; \gamma) = \sum_{\mathbf{m}' \in \mathbf{Z}^2} a(\mathbf{m}'; \gamma)b(\mathbf{m} - \mathbf{m}'; \gamma)\frac{2}{\gamma}\sin\left(\frac{\gamma}{2}\mathbf{m}' \wedge (\mathbf{m} - \mathbf{m}')\right) , \qquad (44)$$

where we set $(\sin x)/x = 1$ for $x = 0$. In particular that for $\gamma = 0$, it coincides with the usual Poisson bracket, namely :

$$\{a, b\}_{\mathrm{cl}} = \{a_{\mathrm{cl}}, b_{\mathrm{cl}}\} = \partial_\theta a_{\mathrm{cl}}\partial_A b_{\mathrm{cl}} - \partial_A a_{\mathrm{cl}}\partial_\theta b_{\mathrm{cl}} , \qquad (45)$$

From (44), the right-hand-side defines a continuous function of γ on I, so that the Poisson bracket $\{a, b\}$ still belongs to $\mathcal{P}_I$. The "Liouville operator" associated to $w \in \mathcal{P}_I$ is the module map defined by :

$$L_w(a) = \{w, a\} , \ a \in \mathcal{P}_I . \qquad (46)$$

The properties of this operator are the following :

$$
\begin{array}{lll}
\text{(i)} & L_w \text{ is } \mathcal{C}(I)-\text{linear} & \\
\text{(ii)} & L_w(a^*) = L_{w^*}(a)^* & w, a \in \mathcal{P}_I , \\
\text{(iii)} & L_w(ab) = L_w(a)b + aL_w(b) & w, a, b \in \mathcal{P}_I , \\
\text{(iv)} & [L_w, L_{w'}] = L_{\{w,w'\}} \text{ (Jacobi's identity)} & w, w' \in \mathcal{P}_I .
\end{array}
\qquad (47)
$$

We also remark that

$$\tau\left(\rho_{\theta,A}(a)\right) = \tau(a) \quad \tau(\{a, b\}) = 0 \quad a, b \in \mathcal{P}_I , \ (\theta, A) \in \mathbf{T}^2 , \qquad (48)$$

which is equivalent to the "integration by parts formula" :

$$\tau(\partial_\mu a \cdot b) = -\tau(a \cdot \partial_\mu b) \qquad \tau\left(L_w(a) \cdot b\right) = -\tau\left(a \cdot L_w(b)\right) , \qquad (49)$$

2.3 The Rotation Algebra $\mathcal{A}_I$

In order to get all continuous functions on our non commutative torus, we ought to define the non commutative analog of the uniform topology on $\mathcal{P}_I$. This can be done by remarking that in the commutative case, the uniform topology is defined through a C^*-norm, namely a norm on the algebra which satisfies :

$$\|ab\| \le \|a\|\|b\| \qquad \|a^*a\| = \|a\|^2 . \qquad (50)$$

The importance of this relation comes from the fact that such a norm is actually entirely defined by the algebraic structure, namely it is given by the spectral radius of a^*a. Therefore, the algebraic structure is sufficient and the uniform topology becomes natural.

To construct such a norm, one uses the representations of $\mathcal{P}_I$. A "representation" of $\mathcal{P}_I$ is a pair $(\pi, \mathcal{H}_\pi)$, where $\mathcal{H}_\pi$ is a separable Hilbert space, and π is a $*$-homomorphism from $\mathcal{P}_I$ into the algebra $\mathcal{B}(\mathcal{H}_\pi)$ of bounded linear operators on $\mathcal{H}_\pi$. The formulæ (17)&(18) give an example of representation for which $\mathcal{H}_\pi = L^2(\mathbf{T}, d\theta/2\pi)$. In particular $\pi(U)$, $\pi(V)$ will be unitary operators on $\mathcal{H}_\pi$ so that if $a \in \mathcal{P}_I$, one gets (if $\|f\|_I$ denotes the sup norm in $\mathcal{C}(I)$) :

$$\|\pi(a)\| \le \sum_{\mathbf{m} \in \mathbf{Z}^2} \|a(\mathbf{m})\|_I < \infty . \tag{51}$$

Two representations $(\pi, \mathcal{H}_\pi)$ and $(\pi', \mathcal{H}_{\pi'})$ are equivalent whenever there is a unitary operator S from $\mathcal{H}_\pi$ into $\mathcal{H}_{\pi'}$ such that for every $a \in \mathcal{P}_I$:

$$S\pi(a)S^{-1} = \pi'(a) . \tag{52}$$

Up to unitary equivalence, one can always assume that $\mathcal{H}_\pi = \ell^2(\mathbf{N})$, so that the family of all equivalence classes of representations of $\mathcal{P}_I$ is a set denoted by $\mathrm{Rep}(\mathcal{P}_I)$. We remark that the norm $\|\pi(a)\|$ depends only upon the equivalence class of π. We then define a seminorm on $\mathcal{P}_I$ by :

$$\|a\|_I = \sup\{\|\pi(a)\|; \pi \in \mathrm{Rep}(\mathcal{P}_I)\} . \tag{53}$$

This notation agrees with the sup-norm on $\mathcal{C}(I)$ if $a \in \mathcal{C}(I)$. Then one has [BaBeFl] :

Proposition 1 *The mapping $a \in \mathcal{P}_I \mapsto \|a\|_I \in \mathbf{R}_+$ is a C^*-norm.*

Remark : The only non trivial fact in this statement is that it is a norm, namely that $\|a\|_I = 0$ implies $a = 0$.

Definition 1 *The algebra $\mathcal{A}_I$ (resp. $\mathcal{A}$) is the completion of $\mathcal{P}_I$ (resp. $\mathcal{P}$) under the norm $\| \cdot \|_I$ (resp. $\| \cdot \|_{\mathbf{T}}$). $\mathcal{A}$ is called the "universal rotation algebra".*

Proposition 2 *1)-Any representation of $\mathcal{P}_I$ extends in a unique way to a representation of $\mathcal{A}_I$*
2)-If $\mathcal{B}$ is any C^*-algebra, and β is a $*$-homomorphism from $\mathcal{P}_I$ to $\mathcal{B}$, then β extends in a unique way as a $*$-homomorphism from $\mathcal{A}_I$ to $\mathcal{B}$.
3)-Any pointwise continuous group of $*$-automorphisms of $\mathcal{P}_I$ extends in a unique way as a norm pointwise continuous group of $*$-automorphisms of $\mathcal{A}_I$.
4)-The trace τ and the angle average $\langle \cdot \rangle$ satisfy :

$$\|\tau(a)\|_I \le \|a\|_I \quad \|\langle a \rangle\|_I \le \|a\|_I \quad a \in \mathcal{P}_I , \tag{54}$$

and therefore they extend uniquely to $\mathcal{A}_I$.
5)-The norm $\| \cdot \|_I$ satisfies :

$$\|a\|_I = \sup_{\gamma \in I} \|\eta_\gamma(a)\| \quad a \in \mathcal{P}_I . \tag{55}$$

In practice the explicit computation of the norm does not require the knowledge of every representation. It is enough to have a faithfull family, namely a family $\{\pi_j\}_{j \in J}$ where J is a set of indices, such that $\pi_j(a) = 0$ for all j's implies $a = 0$. In other words $\cap_{j \in J} \text{Ker}(\pi_j) = \{0\}$. We recall that the spectrum $\text{Sp}(a)$ of an element a of a C^*-algebra with unit $\mathcal{A}$, is the set of complex numbers z such that $z\mathbf{I} - a$ is non invertible in $\mathcal{A}$.

Proposition 3 Let $(\pi_j)_{j \in J}$ be a faithfull family of representations of the C^*-algebra $\mathcal{A}$, then :

$$\|a\|_I = \sup_{j \in J} \|\pi_j(a)\| \quad \text{Sp}(a) = \text{closure}\{\cup_{j \in J} \text{Sp}(\pi_j(a))\} \ . \tag{56}$$

In particular if π is faithfull (namely if J contains only one point), $\|a\|_I = \|\pi(a)\|$ and $\text{Sp}(a) = \text{Sp}(\pi(a))$.

2.4 Smooth functions in $\mathcal{A}_I$

Beside $\mathcal{P}_I$, one can define many dense subalgebras of $\mathcal{A}_I$ playing the role of various subspaces of smooth functions.
(i) For $N \in \mathbf{N}$, the algebra $C^N(\mathcal{A}_I)$ of N-times differentiable elements of $\mathcal{P}_I$ is the completion of $\mathcal{A}_I$ under the norm :

$$\|a\|_{C^N,I} = \sum_{0 \leq n,n';n+n' \leq N} \frac{1}{n!} \frac{1}{n'!} \|\partial_\theta^n \partial_A^{n'}(a)\|_I \ . \tag{57}$$

(ii) $C^\infty(\mathcal{A}_I) = \cap_{N \geq 0} C^N(\mathcal{A}_I)$. It coincides with the set of elements $a = (a(\mathbf{m}))_{\mathbf{m} \in \mathbf{Z}^2}$ with rapidly decreasing Fourier coefficients. It is a nuclear space, similar to the Schwartz space on the torus. Its dual space $S(\mathcal{A}_I)$ is a space of non commutative tempered distributions which can be very useful in investigating unbounded elements.
(iii) $\mathcal{H}^s(\mathcal{A}_I)$ is the Sobolev space, namely the completion of $\mathcal{P}_I$ under the Sobolev norm :

$$\|a\|_{\mathcal{H}^s,I} = \left(\tau(a^*a) + \tau(a^*(-\Delta)^{s/2}a)\right)^{1/2} \quad \Delta = \partial_\theta^2 + \partial_A^2 \ , \tag{58}$$

where $-\Delta$ is the Laplacean on the non commutative torus. The imbedding $\mathcal{H}^{s'}(\mathcal{A}_I) \mapsto \mathcal{H}^s(\mathcal{A}_I)$ is compact if $s' > s$ and $C^\infty(\mathcal{A}_I) = \cap_{s \geq 0} \mathcal{H}^s(\mathcal{A}_I)$, showing that $C^\infty(\mathcal{A}_I)$ is a nuclear space.
(iv) An element of $\mathcal{A}_I$ is holomorphic in some domain D of $(\mathbf{T} + i\mathbf{R})^2$ if the continuous mapping $(\theta, A) \in \mathbf{T}^2 \mapsto \rho_{\theta,A}(a) \in \mathcal{A}_I$, can be extended as a holomorphic function on D. A special interesting case consists in considering the algebra $\mathcal{A}_I(r)$ for $r > 0$, obtained by completing $\mathcal{P}_I$ with the norm :

$$\|a\|_{r,I} = \sup_{\gamma \in I} \sum_{\mathbf{m} \in \mathbf{Z}^2} |a(\mathbf{m}; \gamma)| e^{r|\mathbf{m}|_1} \ , \tag{59}$$

where $|\mathbf{m}|_1 = |m_1| + |m_2|$. Then $\mathcal{A}_I(r)$ becomes a Banach $*$-algebra of holomorphic elements in the strip $D(r) = \{|\text{Im}\theta| < r \ , |\text{Im}A| < r\}$.
(v) Let us consider now the case for which I is an open interval, and let $\mathcal{P}_I^\infty$ be the subalgebra of $\mathcal{P}_I$ the elements of which have Fourier coefficients given by C^∞-functions on I. Let us define the operator ∂_γ on $\mathcal{P}_I^\infty$ by :

$$\partial_\gamma a = \left(\frac{\partial a(\mathbf{m})}{\partial \gamma}\right)_{\mathbf{m} \in \mathbf{Z}^2} \ . \tag{60}$$

Then ∂_γ obeys the following rules (Ito's derivative) :

$$
\begin{array}{lll}
\text{(i)} & \text{it is linear} & \\
\text{(ii)} & \partial_\gamma(a^*) = (\partial_\gamma a)^* & a \in \mathcal{P}_I^\infty , \\
\text{(iii)} & d\tau(a)/d\gamma = \tau(\partial_\gamma a) & a \in \mathcal{P}_I^\infty , \\
\text{(iv)} & \partial_\gamma(ab) = (\partial_\gamma a)b + a(\partial_\gamma b) + \frac{i}{2}(\partial_\theta a \partial_A b - \partial_A a \partial_\theta b) & a, b \in \mathcal{P}_I^\infty .
\end{array}
\tag{61}
$$

One can extend ∂_γ to the dense subalgebra $\mathcal{C}^{N,L}(\mathcal{A}_I)$, obtained by completing $\mathcal{P}_I^\infty$ with respect to the norm :

$$\|a\|_{C^{N,L},I} = \text{Max}_{l \leq L} \|\partial_\gamma^l a\|_{C^N,I} .$$

Let $\| \cdot \|$ be an algebraic $*$-norm. Then the following norm is also an algebraic $*$-norm.

$$\|a\|_{C^1} = \|a\|_I + \|\partial_\theta a\| + \|\partial_A a\| + \|\partial_\gamma a\| .\tag{62}$$

By recursion we will set $\| \cdot \|_{C^N} = (\| \cdot \|_{C^{N-1}})_{C^1}$ with $\| \cdot \|_{C^0} = \| \cdot \|$. It defines then an algebraic $*$-norm on $\mathcal{C}^{N,N}(\mathcal{A}_I)$ equivalent to $\| \cdot \|_{C^{N,N}}$

3 Continuity with respect to Planck's constant

Since the effective Planck constant γ is a tunable physical parameter in many examples, one can wonder whether the various quantities of interest such as mean values of observables, or the evolution, or the spectrum of observables, are continuous functions of γ. The main difficulty in dealing with this problem is that the family of algebras $\gamma \mapsto \mathcal{A}_\gamma$ even though continuous in the sense of Tomiyama [Tom], is not locally trivial. Indeed, $\mathcal{A}_\gamma$ is isomorphic to $\mathcal{A}_{\gamma'}$ if and only if $\gamma = \pm\gamma'$ mod 2π [Rie, PiVo]. Therefore such continuity properties must be carefully studied.

We will give in this section and again without proofs, three kinds of continuity properties. The first concerns the mean value of observables, namely the function $\tau(a)$ if $a \in \mathcal{A}_\gamma$. One important consequence is the Weyl formula for the semiclassical limit of the density of states. The second type of result concerns the continuity of the evolution. It requires the use of a non commutative analog of the Cauchy-Kovaleskaya theorem. In particular, the semiclassical limit of any time correlation function at fixed time, is equal to the corresponding classical expression. The last type of result is the continuity of the gap edges of the spectrum of any observable. This fact will permit to compute the spectrum numerically (see section 4).

It is to be noticed that the algebra $\mathcal{A}_\gamma$ can be constructed from the algebra of pseudodifferential operators of order zero acting on the unit circle. These results are well known in the context of pseudodifferential calculus. However, it turns out that all proofs given here are purely algebraic, and do not require any explicit reference to the form of the symbol. In particular they are valid for any element in the norm closure. But the closure contains much more than pseudodifferential operators, it also contains Fourier integral operators, and elements with no special behaviour.

3.1 Mean values of observables

Our first result is elementary in view of the definition of the algebra $\mathcal{A}_I$.

Proposition 4 *If $a \in \mathcal{A}_I$, the mapping $\gamma \in I \mapsto \tau_\gamma(a)$ is continuous.*

Proof : If $a \in \mathcal{P}_I$ the result comes from the definition of $\mathcal{P}_I$. If $a \in \mathcal{A}_I$, given $\epsilon > 0$, there is $a_\epsilon \in \mathcal{P}_I$ such that $\|a - a_\epsilon\|_I \leq \epsilon$, and therefore by (54), $\sup_{\gamma \in I} |\tau_\gamma(a) - \tau_\gamma(a_\epsilon)| \leq \epsilon$. Thus $\tau(a)$ is a uniform limit of continuous function on I namely it is continuous.

Let us now consider a self adjoint element $H = H^*$ in $\mathcal{A}_I$ and let Σ be its spectrum. Let f be a continuous function on Σ. Then the map $f \in \mathcal{C}(\Sigma) \mapsto \tau_\gamma(f(H)) \in \mathcal{C}$ is linear positive and bounded. Therefore there is a Radon measure $\mathcal{N}_\gamma$ on $\mathbf{R}$ supported by Σ such that :

$$\tau_\gamma(f(H)) = \int_{\mathbf{R}} d\mathcal{N}_\gamma(E) f(E) , \tag{63}$$

This measure is called the "density of states" of H. The "integrated density of states" (IDS) is :

$$\mathcal{N}_\gamma(E) = \int_{E' \leq E} d\mathcal{N}_\gamma(E') . \tag{64}$$

It is a non decreasing function of $E \in \mathbf{R}$. From the proposition 4, we get [BaBeFl] :

Proposition 5 *If $H = H^* \in \mathcal{A}_I$ let $\mathcal{N}_\gamma$ be the integrated density of states of H. Then if E is a point of continuity of $\mathcal{N}_\gamma$, we get :*

$$\lim_{\gamma' \mapsto \gamma} \mathcal{N}_{\gamma'}(E) = \mathcal{N}_\gamma(E) , \tag{65}$$

If I contains $\gamma = 0$, since $\mathcal{A}_0 = \mathcal{C}(\mathbf{T}^2)$ it is easy to check that :

$$\mathcal{N}_0(E) = \int_{-\infty}^{E} d\mathcal{N}_0(E') = \int_{H_{cl}(\theta, A) \leq E} \frac{d\theta dA}{4\pi^2} . \tag{66}$$

where H_{cl} is the Fourier transform of $\eta_0(H)$. Thus $\mathcal{N}_0(E)$ is the area of the set $H_{cl}^{-1}(-\infty, E)$ in the 2-torus. A consequence of the Proposition 4 is the following :

Corollary 2 *If I contains $\gamma = 0$ and $H = H^* \in \mathcal{A}_I$ let $\mathcal{N}_\gamma$ be the integrated density of states of H. Then if E is a real number such that the level set $H_{cl}^{-1}(E)$ has zero Lebesgue measure in the 2-torus, we get :*

$$\lim_{\gamma \to 0} \mathcal{N}_\gamma(E) = \mathcal{N}_0(E) , \text{ (Weyl's formula) } . \tag{67}$$

Let us also mention the following non trivial result [BaBeFl] :

Proposition 6 *If $H = H^* \in \mathcal{P}_I$, then its integrated density of states $\mathcal{N}_\gamma$ is continuous with respect to E for any $\gamma \in I$.*

3.2 The time evolution

Our next result concerns the continuity of the evolution with respect to γ. Let $w \in \mathcal{P}_I$, and let us consider the automorphism of $\mathcal{A}_\gamma$ (for $\gamma \neq 0$) given by :

$$\beta_\gamma(a) = e^{-i\eta_\gamma(w)/\gamma} a e^{i\eta_\gamma(w)/\gamma} , \tag{68}$$

Is it possible to prove that β can be continued at $\gamma = 0$ in such a way as to define an automorphism of $\mathcal{A}_I$? To show that it is actually possible, let us consider the algebra $\mathcal{A}_I(r)$ introduced in 2.4 with the norm defined by (58). Then we get [BeVi]:

Theorem 2 *Let $w = w^*$ be an element of $\mathcal{A}_I(r)$ where I is an interval containing $\gamma = 0$. Then for any ρ such that $0 < \rho < r$,*
(i) the Liouville operator L_w associated to w is well defined as a bounded linear operator from $\mathcal{A}_I(r)$ into $\mathcal{A}_I(r - \rho)$,
(ii) for t small enough, $\exp(tL_w)$ defines a linear bounded operator from $\mathcal{A}_I(r)$ into $\mathcal{A}_I(r - \rho)$,
(iii) $\exp(tL_w)$ can be extended as a $*$-automorphism of $\mathcal{A}_I$ for any $t \in \mathbf{R}$, in such a way as to satisfy :

$$\frac{d e^{tL_w}(a)}{dt} = e^{tL_w}(L_w(a)) , \tag{69}$$

for any $a \in \mathcal{A}_I(r)$.

To prove this result, we will proceed in several steps. First of all :

Lemma 2 *If $w \in \mathcal{A}_I(r_0)$ and $a \in \mathcal{A}_I(r)$, $(r < r_0)$, then for any ρ such that $0 < \rho < r$ one has :*

$$\|\{w, a\}\|_{r-\rho} \leq \frac{2\|w\|_{r_0} \|a\|_r}{e^2 \rho (r_0 - r - \rho)} . \tag{70}$$

Proof : From (44), using the inequalities $|\sin(x)| \leq |x|$, $|\mathbf{m}| \leq |\mathbf{m}'| + |\mathbf{m} - \mathbf{m}'|$ whenever $\mathbf{m}, \mathbf{m}' \in \mathbf{Z}^2$ and $|\mathbf{m}' \wedge \mathbf{m}''| \leq |m_1'\|m''_2| + |m_2'\|m''_1|$ we get :

$$\|\{w, a\}\|_{r-\rho}\| \leq \sup_{\gamma \in I} \sum_{\mathbf{m}', \mathbf{m}'' \in \mathbf{Z}^2} e^{r_0|\mathbf{m}|} |w(\mathbf{m}', \gamma)| e^{r\mathbf{m}''} |a(\mathbf{m}'', \gamma)$$
$$\dots e^{-(r_0 - r + \rho)|\mathbf{m}'| - \rho|\mathbf{m}''|} (|m_1'\|m''_2| + |m_2'\|m''_1|) . \tag{71}$$

The inequality (70) will be obtained by using the estimate $\sup_{n \in \mathbf{Z}} |n| e^{-\rho|n|} = 1/e\rho$.

Lemma 3 *If $w \in \mathcal{A}_I(r)$ for any r such that $0 < \rho < r$ and any $n \in \mathbf{N}$ one has :*

$$\left\|\frac{L_w^n}{n!}\right\|_{r \mapsto r-\rho} \leq \left(\frac{2\|w\|_r}{\rho^2}\right)^n . \tag{72}$$

Proof : One can write

$$\left\|\frac{L_w^n}{n!}\right\|_{r \mapsto r-\rho} \leq \frac{1}{n!} \prod_{k=1}^n \|L_w\|_{r-\rho_{k-1} \mapsto r-\rho_k} ,$$

for any family $(\rho_k)_{1 \leq k \leq n}$ such that $\rho_0 = 0 < \rho_1 < \dots < \rho_{n-1} < \rho_n = \rho$. Using the inequality (70), we get :

$$\left\|\frac{L_w^n}{n!}\right\|_{r \mapsto r-\rho} \le \frac{1}{n!}\left(\frac{2}{e^2}\right)^n \|w\|_r^n \prod_{k=1}^{n} \frac{1}{\rho_k(\rho_k - \rho_{k-1})} \ .$$

Let us choose $\rho_k = \rho k/n$; since $n^n e^{-n} \le n!$, we immediately get the result.

Proof of theorem 2 : the point (i) is exactly the content of the Lemma 2. To prove (ii), it follows from the Lemma 3 that if $|t| < \rho^2/(2\|w\|_r) = T$, the expansion for $\exp(tL_w)$ in powers of t converges in norm as an operator from $\mathcal{A}_I(r)$ into $\mathcal{A}_I(r - \rho)$ and in addition :

$$\|e^{tL_w}\|_{r \mapsto r-\rho} \le \frac{1}{1 - 2|t|\|w\|_r/\rho^2} \ . \tag{73}$$

Proving (iii) is more subtle : if we set $\beta_t = e^{tL_w}$, we observe that since L_w is a $*$-derivation, by (47), then, if $a, b \in \mathcal{A}_I(r)$:

$$
\begin{array}{lll}
\text{(i)} & \beta_t(ab) = \beta_t(a)\beta_t(b) & \text{for } |t| < T \ , \\
\text{(ii)} & \beta_t(a^*) = \beta_t(a)^* & \text{for } |t| < T \ , \\
\text{(iii)} & \beta_{t+s}(a) = \beta_t(\beta_s(a)) & \text{for } |t| + |s| < T \ , \\
\text{(iv)} & d\beta_t(a)/dt = \beta_t(L_w(a)) & \text{for } |t| < T \ .
\end{array}
\tag{74}
$$

Therefore given any representation π of $\mathcal{P}_I$, π can be extended as a representation of $\mathcal{A}_I$ thus of $\mathcal{A}_I(r)$. In particular, $\pi \circ \beta_t$ gives also a representation of $\mathcal{A}_I(r)$, so that by the same type of argument used in 2.3 (see (51)&(53)), one gets $\|\pi \circ \beta_t(a)\|_I \le \|a\|_I$, and since π is arbitrary :

$$\|\beta_t(a)\|_I \le \|a\|_I \ , \text{ for } |t| < T. \tag{75}$$

In particular, β_t can be extended by continuity to $\mathcal{A}_I$ and the extension still satisfies (74). Now if $t \in \mathbf{R}$, let n be a positive integer such that $|t/n| < T$. Then we set $\beta_t = (\beta_{t/n})^n$. Thanks to (74) (iii), it is standard to check that this definition does not depend upon the choice of n. Moreover, (74) continues to hold at any value of t : this is obvious for (i), (ii), (iii). (iv) also holds once one notices that L_w commutes with β_t. Therefore $(\beta_t)_{t \in \mathbf{R}}$ defines a 1-parameter group of $*$-automorphisms of $\mathcal{A}_I$. At last it is norm-pointwise continuous, namely :

$$\lim_{t \to 0} \|\beta_t(a) - a\|_I = 0 \ , \tag{76}$$

For indeed, by a 3ϵ-argument, it is enough to check it for $a \in \mathcal{A}_I(r)$, which is simply a consequence of the Lemma 3.

3.3   The spectrum of observables

Our last result concerns the continuity of the spectrum with respect to γ. Let $(\Sigma(t))_{t \in \mathbf{R}}$ be a family of compact subsets of a topological space X. This family is continuous at $t = t_0$ if the two following properties hold :

(i) it is continuous from the outside, namely given any closed set F in X, such that $\Sigma(t_0) \cap F = \emptyset$, there i $\delta > 0$, such that if $|t - t_0| \le \delta$, then $\Sigma(t) \cap F = \emptyset$.

(ii) it is continuous from inside, namely given any open set O in X, such that $\Sigma(t_0) \cap O \neq \emptyset$, there is $\delta > 0$, such that if $|t - t_0| \leq \delta$, then $\Sigma(t) \cap O \neq \emptyset$.

If $X = \mathbf{R}$ a gap of $\Sigma(t)$ is a connected component of $\mathbf{R} - \Sigma(t)$. One can check that this definition is equivalent to the continuity of the gap edges of $\Sigma(t)$ at t_0.

For $a \in \mathcal{A}_I$ we set $\Sigma(\gamma) = \mathrm{Sp}(\eta_\gamma(a))$, whenever $\gamma \in I$. The main result of this subsection is [BaBeFl] :

Theorem 3 *For any normal element $a \in \mathcal{A}_I$, (namely such that $aa^* = a^*a$), the family $(\Sigma(\gamma))_{\gamma \in I}$ is continuous at every point of I.*

The proof of this theorem will not be given here. It can be found in [BaBeFl]. However, it is of very high importance in view of the numerical computation of the spectrum. For we will see in the next section that the spectrum can be easily computed on a computer for rational values of $\gamma/2\pi$. The continuity of the gap edges everywhere on I implies that this type of computation is sufficient to get an idea of the spectrum for irrational values of $\gamma/2\pi$.

Actually for smooth self adjoint elements of $\mathcal{A}_I$ one gets a better result [BaBeFl], namely :

Theorem 4 *For any self adjoint element $H \in C^{3,1}\mathcal{A}_I$, the gap edges of any open gap of $(\Sigma(\gamma))_{\gamma \in I}$ are Lipshitz continuous at every point of I.*

Similar but weaker results have already been obtained previously by Choi et al. [ChElYu], and by Avron et al. [AvSi] on the almost Mathieu model. They found Hölder continuity only. Here we get a stronger result. However the Lipshitz constant depends explicitly of the width of the gap considered, and it diverges whenever the width tends to zero.

As we will see in section 4 below, there is no chance to get a better result because the gap edges have discontinuous derivative at each rational value of $\gamma/2\pi$. On the other hand, if a gap closes for some value of γ then generically with respect to $H \in C^{3,1}\mathcal{A}_I$, we only get Hölder continuity with exponent $1/2$ near this point.

4 Structure of the Rotation Algebra $\mathcal{A}_I$

In this section we will give without proofs, a description of the structure of the rotation algebra. The reader interested in the proofs will be refered to [BaBeFl].

Let us consider first the case $\gamma = 0$. The algebra $\mathcal{P}_0$ is then the convolution algebra associated to the group $\mathbf{Z}^2$. Therefore by Fourier transform, one transforms it into the algebra of trigonometric polynomials with the pointwise multiplication. More precisely, if $a \in \mathcal{P}_0$, we set :

$$a_{\mathrm{cl}}(\theta, A) = \sum_{\mathbf{m} \in \mathbf{Z}^2} a(\mathbf{m}) e^{i(\theta m_1 - A m_2)} \ . \tag{77}$$

This a trigonometric polynomial. The main properties of the Fourier transform are :

$$(ab)_{\mathrm{cl}}(\theta, A) = a_{\mathrm{cl}}(\theta, A) b_{\mathrm{cl}}(\theta, A) \ , \ a, b \in \mathcal{P}_0 \ , \tag{78}$$

and

$$(a^*)_{\mathrm{cl}}(\theta, A) = a_{\mathrm{cl}}(\theta, A)^* \ , \ a \in \mathcal{P}_0 \ . \tag{79}$$

It follows that for every $(\theta, A) \in \mathbf{T}^2$, the map $a \in \mathcal{P}_0 \mapsto a_{\mathrm{cl}}(\theta, A) \in \mathbf{C}$ is a representation. Therefore :

$$\sup_{(\theta, A) \in \mathbf{T}^2} |a_{\mathrm{cl}}(\theta, A)| \leq \|a\|_0 \ , \ a \in \mathcal{P}_0 \ . \tag{80}$$

In particular, the Fourier transform $a \in \mathcal{P}_0 \mapsto a_{\mathrm{cl}} \in \mathcal{C}(\mathbf{T}^2)$, extends to $\mathcal{A}_0$ as a $*$-homomorphism. As a consequence of the Gelfand theorem we get the following result [BaBeFl] :

Theorem 5 *The Fourier transform $a \in \mathcal{P}_0 \mapsto a_{\mathrm{cl}} \in \mathcal{C}(\mathbf{T}^2)$, extends as a $*$-isomorphism from $\mathcal{A}_0$ to $\mathcal{C}(\mathbf{T}^2)$.*

Let us now consider the case $\gamma = 2\pi p/q$ where p, q are positive integers prime to each others. As we have seen in section 2, $\mathcal{A}_\gamma$ is isomorphic to $\mathcal{A}_{\gamma+2\pi}$, so that one can assume that $0 < p < q$ without loss of generality. One can extend the previous analysis to this case by introducing two $q \times q$ unitary matrices u and v satisfying :

$$u^q = \mathbf{I} = v^q \ , \qquad uv = e^{2i\pi p/q} vu \ . \tag{81}$$

The Fourier transform of $a \in \mathcal{P}_\gamma$ is then given by the following matrix valued function :

$$a_{\mathrm{cl}}(\theta, A) = \sum_{\mathbf{m} \in \mathbf{Z}^2} a(\mathbf{m}) e^{i(\theta m_1 - A m_2)} w(\mathbf{m}) \ , \tag{82}$$

where :

$$w(\mathbf{m}) = u^{m_1} v^{m_2} e^{-i\pi p m_1 m_2 / q} \ . \tag{83}$$

We remark that in this last expression, $w(\mathbf{m} + q\mathbf{m}') = w(\mathbf{m})$, namely $\mathbf{m}$ is defined modulo q.

An example of such pair of matrices is given by :

$$u = \begin{bmatrix} 0 & 1 & 0 & \cdots & 0 & 0 \\ 0 & 0 & 1 & \cdots & 0 & 0 \\ \cdots & \cdots & \cdots & \cdots & \cdots & \cdots \\ \cdots & \cdots & \cdots & \cdots & \cdots & \cdots \\ 0 & 0 & 0 & \cdots & 0 & 1 \\ 1 & 0 & 0 & \cdots & 0 & 0 \end{bmatrix} \ , \ v = \begin{bmatrix} 1 & 0 & 0 & \cdots & 0 & 0 \\ 0 & \lambda & 0 & \cdots & 0 & 0 \\ \cdots & \cdots & \cdots & \cdots & \cdots & \cdots \\ \cdots & \cdots & \cdots & \cdots & \cdots & \cdots \\ 0 & 0 & 0 & \cdots & \lambda^{q-2} & 0 \\ 0 & 0 & 0 & \cdots & 0 & \lambda^{q-1} \end{bmatrix} \ , \tag{84}$$

where $\lambda = e^{2i\pi p/q}$. Actually, any such pair is unitarily equivalent to this latter one. Let us also set :

$$w'(m_1, m_2) = w(-m_2, m_1) \ , \tag{85}$$

to get the following characterization of $\mathcal{A}_{2\pi p/q}$ [BaBeFl] :

Theorem 6 *The Fourier transform $a \in \mathcal{P}_{2\pi p/q} \mapsto a_{\mathrm{cl}} \in \mathcal{C}(\mathbf{T}^2) \otimes M_q$, extends as a $*$-isomorphism from $\mathcal{A}_{2\pi p/q}$ to the subalgebra $\mathcal{C}_{\mathrm{cov}}(\mathbf{T}^2, q)$ of $\mathcal{C}(\mathbf{T}^2) \otimes M_q$, the element of which being continuous functions a_{cl} from $\mathbf{T}^2$ into M_q satisfying the covariance condition*

$$w'(\mathbf{m}) a_{\mathrm{cl}}(\theta, A) w'(\mathbf{m})^{-1} = a_{\mathrm{cl}}((\theta, A) + 2\pi \mathbf{m} p/q) \ . \tag{86}$$

The main interest of this result is that it makes it possible to compute the spectrum of an element $a \in \mathcal{A}_{2\pi p/q}$. For indeed for every $(\theta, A) \in \mathbf{T}^2$, the map $\pi(\theta, A) : a \in \mathcal{A}_{2\pi p/q} \mapsto a_{\mathrm{cl}}(\theta, A) \in M_q$ is a representation, and the family $\{\pi(\theta, A); (\theta, A) \in \mathbf{T}^2\}$ is faithfull. Therefore, thanks to prop.3, denoting by $e_k(\theta, A)$, $1 \le k \le q$, the eigenvalues of $a_{\mathrm{cl}}(\theta, A)$, the spectrum of a is :

$$\mathrm{Sp}(a) = \cup_{1 \le k \le q} \mathrm{Im}(e_k) \ , \ \mathrm{Im}(e_k) = \{e_k(\theta, A) \in \mathbf{C}; (\theta, A) \in \mathbf{T}^2\} \ . \tag{87}$$

Each set $\mathrm{Im}(e_k)$ is called a "band". The computation of the eigenvalues can be done numerically by matrix diagonalization. In many important examples, such as the "Harper" model (see Fig. 1) given by :

$$H_{\mathrm{Harper}} = U + U^* + V + V^* \ , \tag{88}$$

it is possible to compute analytically the points in the 2-torus for which the band edges are reached. In these cases, the numerical computation of the spectrum requires to diagonalize only few matrices for each value of p/q [BaBeFl].

The theorems 4 and 5 can be rephrased by characterizing the set of (closed two-sided) ideals of $\mathcal{A}_{2\pi p/q}$. If $p = 0$, any ideal J is given by the space of continuous functions on $\mathbf{T}^2$ vanishing on some closed subset Ω_J of $\mathbf{T}^2$. The map $J \mapsto \Omega_J$ is actually one-to-one. If $p \ne 0$, the same is true if we demand that Ω_J be invariant by the translations of period $2\pi/q$ in the 2-torus [BaBeFl]. However If $\gamma/2\pi$ is irrational, we get the following result [Sla, BaBeFl] :

Theorem 7 *If $\gamma/2\pi$ is irrational the algebra $\mathcal{A}_\gamma$ is simple, namely there is no other ideal than $\{0\}$ and $\mathcal{A}_\gamma$ itself.*

Corollary 3 *If $\gamma/2\pi$ is irrational every representation of the algebra $\mathcal{A}_\gamma$ is faithfull .*

A nice proof of theorem 6 was provided by Slawny [Sla], and the reader will find it in [BaBeFl]. The corollary is an immediate consequence of that theorem, for if π is a representation of the algebra $\mathcal{A}_\gamma$ its kernel is an ideal, namely it is either the algebra $\mathcal{A}_\gamma$, in which case $\pi = 0$, which is not possible since $\pi(\mathbf{I}) = 1$, or it vanishes, namely π is faithfull.

Thanks to this last result, we can choose any representation to produce explicit calculations. The three theorems of this section are sufficient to characterize the algebra $\mathcal{A}_I$ for any compact subset I of $\mathbf{R}$. For indeed thanks to (55) (prop. 2), J is an ideal of $\mathcal{A}_I$ if and only if for any $\gamma \in I$, $\eta_\gamma(J)$ is an ideal of $\mathcal{A}_\gamma$. The ideal structure is sufficient to characterize any C^*-algebra $\mathcal{B}$ which is a homomorphic image of $\mathcal{A}_I$.

5 Semiclassical asymptotics for the spectrum

In this section we will denote by $H = H^*$ a selfadjoint element of $\mathcal{A}_I$, and we intend to given a description of its spectrum when I is a small open interval around $\gamma = 2\pi p/q$. The same kind of results can be obtained for a unitary operator, and this will be left as an exercise to the reader.

5.1 2D lattice electrons in a magnetic field

In this subsection, we will describe a physical situation where the rotation algebra enters as an essential tool. It will give us a different intuitive description of the problem which may be useful.

Let us consider a two dimensional lattice, with lattice spacing δ, that we will identify with $\mathbf{Z}^2$, on which charged particles like electrons or holes, are supposed to move. Their quantum states will be wave functions $\psi = (\psi(\mathbf{m}))_{\mathbf{m} \in \mathbf{Z}^2} \in \ell^2(\mathbf{Z}^2)$. We suppose in addition that a uniform magnetic field B is applied on this lattice, perpendicularly to the plane of the lattice. Let $\mathbf{A} = (A_1, A_2)$ be the corresponding vector potential, namely a vector field on $\mathbf{R}^2$ solution of the equation $\partial_1 A_2 - \partial_2 A_1 = B$. In the Hilbert space $\ell^2(\mathbf{Z}^2)$, we consider the "magnetic translations" T_1, T_2 studied by Zak [Zak], associated to the two basis vectors $e_1 = (1, 0)$ and $e_2 = (0, 1)$ of $\mathbf{Z}^2$, namely the unitary operators defined by :

$$T_\mu \psi(\mathbf{m}) = e^{2i\pi \frac{e}{h} \int_{(\mathbf{m}-e_\mu)\delta}^{\mathbf{m}\delta} d\mathbf{l} \cdot \mathbf{A}} \psi(\mathbf{m} - e_\mu) , \qquad \mu = 1, 2 , \tag{89}$$

where e is the electric charge of the particle, h is the Planck constant and the integral in the phase factor is computed along the segment joining the sites $\mathbf{m} - e_\mu$ and $\mathbf{m}$. These two operators satisfy the following commutation rule :

$$T_1 T_2 = e^{2i\pi\phi/\phi_0} T_2 T_1 , \tag{90}$$

where ϕ is the magnetic flux through the unit cell, and $\phi_0 = h/e$ is the flux quantum. Therefore these two operators generate a representation of the C^*-algebra $\mathcal{A}_\gamma$ where now

$$\gamma = 2\pi\phi/\phi_0 = \text{const}.B , \tag{91}$$

is proportional to the magnetic field. In practice, if the lattice is given by the positions of the ions of a metal, δ is of the order of 1 Å so that even with the highest kind of magnetic field that can be produced in laboratories, namely $B \approx 18$ Teslas , we get $\gamma/2\pi \approx 0.5 \ 10^{-4}$ which is fairly small, and shows that in this situation a semiclassical approximation will always be valid. However during the last ten years, networks with lattice spacings of the order of the micrometer have been built [PaChRa], leading to values of $\gamma/2\pi$ of the order of unity in magnetic fields not larger than 40 Gauss. This is why it has been necessary to go beyond the semiclassical regime.

¿From the band theory of metals [MeAs], the conduction properties are given only by those electrons sitting in the conduction bands, namely with energies within an interval of order $k\Theta$ from the Fermi level, if Θ denotes here the temperature, and k the Boltzmann constant. If we assume for simplicity that there is only one such band, thanks to the so-called "Peierls substitution" [Pei], one can prove rigorously that the restriction H of the Hamiltonian to that band is given by a selfadjoint element of the C^*-algebra generated by the two magnetic translations [Bel:Eva]. If several bands have to be considered, the Hamiltonian will be represented by a matrix with entries in this algebra. For $B = 0$, the band Hamiltonian H is represented through its Fourier transform (see section 4) by a continuous function H_{cl} on $\mathbf{T}^2$. In Solid State Physics, one usually uses the quasimomentum notation $\mathbf{k} = (k_1, k_2)$ instead of (θ, A) to represent a point in this 2-torus. Thus we get the following correspondence :

$$(T_1)_{\text{cl}} = e^{ik_1} , \ (T_2)_{\text{cl}} = e^{ik_2} , \text{ if } B = 0 . \tag{92}$$

The advantage of this latter notation is that it restores the symmetry between the two directions in the lattice, a natural fact in the present context, even though it does not look so natural in the kicked rotor problem. In this subsection, we will prefer the use of the quasimomentum notations instead of the action-angle ones.

Let us now remark that the representation given by (89) is actually a very natural one from algebraic point of view. For indeed, thanks to subsection 2.2 (34&(35), the Hilbert space $\ell^2(\mathbf{Z}^2)$ can be seen as the completion $L^2(\mathcal{A}_\gamma, \tau)$ of the prehilbert space $\mathcal{P}_\gamma$ endowed with the scalar product $\langle a|b \rangle = \tau(a^*b)$. Let η be the natural imbedding of $\mathcal{P}_\gamma$ into $L^2(\mathcal{A}_\gamma, \tau)$. Since $\tau(a^*a) \leq \|a^*a\|_\gamma$, η can be extended to $\mathcal{A}_\gamma$ by continuity. Then let π_{GNS} be the representation of $\mathcal{A}_\gamma$ on this Hilbert space given by the left multiplication, namely

$$\pi_{\text{GNS}}(a)\eta(b) = \eta(ab) , \, a, b \in \mathcal{A}_\gamma . \tag{93}$$

The name "GNS" refers to Gelfand-Naimark-Segal [Dix], who defined and studied this representation in a C^*-algebra. Then we claim the following [BaBeFl]:

Theorem 8 *The representation of $\mathcal{A}_\gamma$ given by the magnetic translations in (89) is unitarily equivalent to the GNS representation relative to the trace of $\mathcal{A}_\gamma$. This representation is faithfull for any values of γ.*

Actually, the representation given by the magnetic translations depends upon the choice of a vector potential. The GNS representation corresponds to the so-called "symmetric gauge", namely $\mathbf{A} = B(-x_2, x_1)$. Every other gauge can be reached by a unitary transformation.

5.2 Low field expansion

We now consider an interval I of the form $I = [-\epsilon_0, \epsilon_0]$, for some $\epsilon_0 > 0$, and let $H = H^*$ belong to $\mathcal{A}_I$. We will describe a semiclassical expansion near a bottom well. In order to do so, let us assume that H_{cl} admits a local minimum or a local maximum at $\mathbf{k} = \mathbf{k}_0$. Moreover we will assume that this extremum is regular which is a generic property. More precisely, and without loss of generality we will assume :

(H0) $H = H^* \in C^N(\mathcal{A}_I)$ for $N > 2$, and all its Fourier coefficients are N-times differentiable with respect to γ.
(H1) H_{cl} admits a local minimum at $\mathbf{k}_0 = (0,0)$ and with $H_{\text{cl}}(0,0) = 0$.
(H2) The Hessian $D^2 H_{\text{cl}}(0,0)$ of H_{cl} at $\mathbf{k} = (0,0)$ is a positive definite 2×2 matrix.

Our goal is to describe the spectrum of $\eta_\gamma(H)$ near the energies E close to $H_{\text{cl}}(0,0) = 0$ for $\gamma \in I$. To describe the result, let us assume that H can be written as :

$$\eta_\gamma(H) = \sum_{\mathbf{m} \in \mathbf{Z}^2} h(\mathbf{m}; \gamma) W_\gamma(\mathbf{m}) . \tag{94}$$

with $h(\mathbf{m}; \gamma)^* = h(-\mathbf{m}; \gamma)$. Since H is smooth, one can check that this series converges absolutely in norm, so that this expansion is meaningfull. Let us introduce the following function :

$$H_{\text{scl}}(\mathbf{k}; \gamma) = \sum_{\mathbf{m} \in \mathbf{Z}^2} h(\mathbf{m}; \gamma) e^{i(m_1 k_1 + m_2 k_2)} , \, \mathbf{k} \in \mathbf{T}^2 , \gamma \in I , \tag{95}$$

which coincides with H_{cl} for $\gamma = 0$. Then we get the following result [Bel:Eva, BaBeFl]

Theorem 9 *Let H satisfy (H0),(H1)&(H2), and let H_{scl} be defined by (95). Then there are $\delta > 0$ and $0 < \epsilon \leq \epsilon_0$ such that if $|\gamma| \leq \epsilon$, the set $\mathrm{Sp}(\eta_\gamma(H)) \cap (-\delta, +\delta)$ is contained in the union over $n \in \mathbf{N}$ of the intervals $J_n(\gamma) = [E_n(\gamma) - \Delta(\gamma), E_n(\gamma) + \Delta(\gamma)]$ where $E_n(\gamma)$ admits a Taylor expansion in γ up to the Nth order of the form :*

$$E_n(\gamma) = |\gamma|\left(n + \frac{1}{2}\right)\left(\det D^2 H_{cl}(0,0)\right)^{1/2} + \gamma\left(\frac{\partial H_{scl}}{\partial \gamma}\right)_{\gamma=0,\mathbf{k}=\mathbf{0}} + \cdots + O(\gamma^N) , \quad (96)$$

$$0 < \Delta(\gamma) \leq \mathrm{const.}|\gamma|^{N'} , \text{ for some } N' > N . \quad (97)$$

To understand more intuitively this result let us introduce a faithfull representation π of $\mathcal{A}_\gamma$ in a Hilbert space in which one can find two selfadjoint operators K_1, K_2 such that $[K_2, K_1] = i\mathrm{sgn}(\gamma)$, where $\mathrm{sgn}(x)$ denotes the sign of the real number x. That such a representation exists is a well known fact [BaBeFl], and is a consequence of the Weyl theorem on the canonical commutation relations. In this representation, one has :

$$\pi(U) = e^{i|\gamma|^{1/2}K_1} , \quad \pi(V) = e^{i|\gamma|^{1/2}K_2} , \quad (98)$$

Therefore we get from (94) :

$$H_\gamma = \pi \circ \eta_\gamma(H) = \sum_{\mathbf{m} \in \mathbf{Z}^2} h(\mathbf{m}; \gamma)e^{i|\gamma|(m_1 K_1 + m_2 K_2)} , \quad \gamma \in I . \quad (99)$$

Let us expand this expression formally in powers of $|\gamma|^{1/2}$ to obtain :

$$H_\gamma = \gamma \partial_\gamma H_{scl}(0; 0) + \frac{1}{2}|\gamma|\partial_\mu \partial_\nu H_{scl}(0,0)K_\mu K_\nu + O\left(|\gamma|^{3/2}\right) , \quad (100)$$

where we have used the Einstein convention on the repeated indices (here $\mu, \nu \in \{1, 2\}$). By a unitary transformation, the quadratic term can be transformed into $\omega(K_1^2 + K_2^2)/2$ where ω is the determinant of the Hessian matrix $\partial_\mu \partial_\nu H_{scl}(0,0)$. We recognize here the Hamiltonian of a harmonic oscillator. Actually, if we choose the representation corresponding to a 2D free electron in a uniform magnetic field, namely the Hilbert space is $L^2(\mathbf{R}^2)$, and $K_\mu = \mathrm{const.}(P_\mu - eA_\mu)$ for some physical constant, then this Hamiltonian is the Landau one, namely the Hamiltonian describing a free electron in a uniform magnetic field. For this reason, the energy levels E_n are called the "Landau levels" and are equal to that order in γ to $\omega(n + 1/2)$, leading to the expression (96).

The proof of this theorem can be found in [BaBeFl, Bel:Eva]. The calculation of E_n to the next order has been done in [RaBe:Alg], in the case for which $\partial_\gamma H = 0$, and leads to (for $\gamma > 0$) :

$$E_n = \gamma\omega(2n + 1)/2 + \gamma^2\Delta^2 H_{cl}(0)\left(1 + (2n + 1)^2\right)/64$$

$$- \cdots \gamma^2\left[9(3n^2 + 3n + 1)|\delta H_{cl}(0)|^2 + (3n^2 + 3n + 2)|\partial^3 H_{cl}(0)|^2\right]/288\omega + O(\gamma^3) , \quad (101)$$

where :

$$\partial = \frac{\partial}{\partial k_1} - i\frac{\partial}{\partial k_2} \ , \ \bar{\partial} = \frac{\partial}{\partial k_1} + i\frac{\partial}{\partial k_2} \ , \ \Delta = \partial\bar{\partial} \ . \tag{102}$$

These formulæ have been checked numerically on several models. The calculation to the third order in powers of γ havsbeen computed for the Harper model (see (88)) [RaBe:Alg] and gives for the minimum (Fig. 1) :

$$E_n = -4 + \gamma(2n+1) - \frac{\gamma^2}{16}\left[1 + (2n+1)^2\right] + \frac{\gamma^3}{192}\left[n^3 + (n+1)^3\right] + O(\gamma^4) \ . \tag{103}$$

Another example of interest has been investigated in [BeKrSe] and concerns the nearest neighbour model on a triangular lattice with two fluxes (Fig. 2). The corresponding Hamiltonian is given by :

$$H_\Delta = T_1 + T_2 + T_3 + T_1^* + T_2^* + T_3^* \ , \text{with} \quad T_1 T_2 T_3 = e^{i2\pi\phi'/\phi_0} \ , \tag{104}$$

and (89). Here ϕ' represents the flux through the "up" triangles, while $\phi - \phi'$ represents the flux through the "down" triangles. The corresponding classical counterpart is given by :

$$H_{\Delta,\mathrm{cl}} = 2\cos k_1 + 2\cos k_2 + 2\cos(k_1 + k_2 + \frac{\gamma}{2} - \gamma') \ , \tag{105}$$

if we set $\gamma' = 2\pi\phi'/\phi_0$. The minima and maxima occur at the points $k_1 = k_2 = 2\pi\sigma/3 + \gamma'/3 = \theta_\sigma$, where $\sigma = -1, 0, +1$. If $\theta_\sigma \neq \pi/2$, one gets the following expression [BeKrSe] :

$$E_{\sigma,n} = 6\cos\theta_\sigma - \gamma\sqrt{3}(2n+1)\cos\theta_\sigma + \gamma\sin\theta_\sigma + \gamma^2\left[1 + (2n+1)^2\right]\cos\theta_\sigma/8 +$$

$$\cdots + \gamma^2\sin 2\theta_\sigma\cos\theta_\sigma[3(2n+1) + 5]/72 - \gamma^2/12\cos\theta_\sigma - \gamma^2\sqrt{3}\sin\theta_\sigma(2n+1)/6 \ , \tag{106}$$

giving rise to three bundles of Landau levels. For $\gamma' \approx 0$ two of these bundles are very close and actually intersect each other (Fig. 3) . The comparison between this formula and the exact spectrum obtained by matrix diagonalization is very good : they agree up to four digits for the coefficients of the power expansion in γ [BeKrSe].

The assumptions (H0, H1, H2) concern the generic case, for which the extremum is regular. However, some non generic case has been observed. For example, in [Wil:Cri, BaKr], the case of a square lattice with second nearest neighbour has been studied. The corresponding Hamiltonian is :

$$H_{\mathrm{WBK}} = T_1 + T_1^* + T_2 + T_2^* + t_2\left(T_1^2 + T_1^{2^*} + T_2^2 + T_2^{2^*}\right) \ . \tag{107}$$

For $t_2 < 1/4$, the classical Hamiltonian has only one absolute minimum, like in the Harper case. At the value $t_2 = 1/4$, this minimum bifurcates to give four degenerate minima for $\epsilon > 1/4$. At the bifurcation value, this minimum becomes flat namely the Hessian actually vanishes identically, giving rise to a normal form like :

$$H_{\mathrm{WBK},\gamma} = -3 + \frac{\gamma^2}{4}\left(K_1^4 + K_2^4\right) + O(\gamma^4) \ . \tag{108}$$

A Bohr-Sommerfeld quantization condition gives at the lowest order in γ :

$$E_n = -3 + \frac{\gamma^2\pi}{4\Gamma(1/4)^4}(2n+1)^2 + O(\gamma^4) \quad \text{for } n \text{ large} \ , \tag{109}$$

giving parabolic Landau levels, as can be observed in (Fig. 4).

5.3 Expansion near a rational field

The method outlined in the previous subsection for a low field expansion of the spectrum, can be extended to the expansion near a rational field, or also to the case of a matrix Hamiltonian, as can happen if several different bands contribute to the conduction. We will give here the method for the rational fields, leaving to the reader the case of a matrix Hamiltonian as an exercise.

We consider now an interval $I = [2\pi p/q - \epsilon_0, 2\pi p/q + \epsilon_0]$ and $H = H^* \in \mathcal{A}_I$. Using the matrices u, v given in section 4 (84), letting U_γ, V_γ be the generators of the algebra $\mathcal{A}_\gamma$ for $|\gamma| \leq \epsilon_0$, we consider the elements U', V' in $\mathcal{A}_\gamma \otimes M_q$ defined by :

$$U' = U_\gamma \otimes u \ , \ V' = V_\gamma \otimes v \ . \tag{110}$$

They are unitary and satisfy the commutation rule :

$$U'V' = e^{i(\gamma + 2\pi p/q)} V'U' \ , \tag{111}$$

showing that they generate in $\mathcal{A}_\gamma \otimes M_q$ a subalgebra $*$-isomorphic to $\mathcal{A}_{\gamma + 2\pi p/q}$. Letting γ vary in $I(0) = [-\epsilon_0, +\epsilon_0]$, we get a $*$-isomorphism between $\mathcal{A}_I$ and a closed subalgebra of $\mathcal{A}_I(0) \otimes M_q$.

Assuming that H is smooth enough, one can expand it as :

$$\eta_{\gamma + 2\pi p/q}(H) = H_\gamma = \sum_{\mathbf{m} \in \mathbf{Z}^2} h(\mathbf{m}; \gamma) W_\gamma(\mathbf{m}) \otimes w(\mathbf{m}) \ , \ \gamma \in I(0) \ , \tag{112}$$

and this series converges in norm. So we are left with the same problem as in 5.2, with now matrix Hamiltonians instead. Following the same scheme, the classical counterpart is the matrix valued function :

$$H_{\mathrm{scl}}(\mathbf{k}; \gamma) = \sum_{\mathbf{m} \in \mathbf{Z}^2} h(\mathbf{m}; \gamma) e^{i(m_1 k_1 + m_2 k_2)} w(\mathbf{m}) \ , \ \mathbf{k} \in \mathbf{T}^2 \ , \ \gamma \in I(0) \ . \tag{113}$$

As we already indicated, we must first diagonalize this matrix at $\gamma = 0$, giving q real eigenvalues (since H is selfadjoint), $e_1(\mathbf{k}), e_2(\mathbf{k}), \cdots, e_q(\mathbf{k})$, and therefore q bands $B_1, B_2, \cdots, B_q$, namely the set of values of the $e_j(\mathbf{k})$'s as $\mathbf{k}$ varies in the 2-torus. Since H is selfadjoint, it is always possible to choose the $e_j(\mathbf{k})$'s smooth. We will also denote by $P_1(\mathbf{k}), P_2(\mathbf{k}), \cdots, P_q(\mathbf{k})$ the corresponding eigenprojections; they are also smooth with respect to $\mathbf{k}$. We will now assume the following :

(H$_q$0) $H = H^* \in \mathcal{C}^N(\mathcal{A}_I)$ for some $N > 2$, and all coefficients in the expansion (113) are N-times continuously differentiable with respect to γ.

(H$_q$1) The eigenvalue e_j admits a minimum at $\mathbf{k} = 0$, and $e_j(0) = 0$. Moreover, no other eigenvalue of $H_{\mathrm{scl}}(0; 0)$ coincides with $e_j(0) = 0$.

(H$_q$2) The minimum of e_j is regular, namely the Hessian matrix $\partial_\mu \partial_\nu e_j(0)$ is positive definite.

Then we get the following result [Bel:Eva, BaBeFl] :

Theorem 10 *Let H satisfy (H$_q$0), (H$_q$1), (H$_q$2). Then there are $\delta > 0$ and $0 < \epsilon \leq \epsilon_0$ such that if $|\gamma| < \epsilon$, the set $\mathrm{Sp}(\eta_{\gamma + 2\pi p/q}(H)) \cap (-\delta, +\delta)$ contains a subset Σ_j which is itself contained in the union over $n \in \mathbf{N}$ of the interval $J_{n,j}(\gamma) = [E_{n,j}(\gamma) -$*

$\Delta(\gamma), E_{n,j}(\gamma) + \Delta(\gamma)]$ where $E_{n,j}(\gamma)$ admits a Taylor expansion in γ up to the Nth order of the form :

$$E_{n,j}(\gamma) = |\gamma|(n + 1/2) \left(\det D^2 e_j(\mathbf{0})\right)^{1/2} + \gamma \left(\frac{\partial e_j}{\partial \gamma}\right)_{\gamma=0, \mathbf{k}=0} + \gamma E_{\text{RW}} + O(\gamma^2) , \quad (114)$$

where $0 < \Delta(\gamma) \leq \text{const.}|\gamma|N'$, for some $N' > N$, where E_{RW} is the "Rammal-Wilkinson" term given by the following expression :

$$E_{\text{RW}} = \frac{i}{2}\text{Tr}\left(P_j(0)[\partial_1 H_{\text{scl}}(0)\partial_2 P_j(0) - \partial_2 H_{\text{scl}}(0)\partial_1 P_j(0)]\right) , \quad (115)$$

The strategy used to prove this theorem is based upon the so-called "Schur complement formula". Let $H = H^*$ be a selfadjoint operator acting on a Hibert space of the form $\mathcal{H} = \mathcal{P} \oplus \mathcal{Q}$. Let P, Q be the orthogonal projections on each subspace of that decomposition and let D be a partial isometry from $\mathcal{H}$ to $\mathcal{P}$ such that $DD^* = I_{\mathcal{P}}$ and $D^*D = P$. We define on $\mathcal{P}$ the family of operators :

$$H_{\text{eff}}(z) = DHD^* + DHQ(zI - QHQ)^{-1}QHD^* , \quad (116)$$

whenever z is a complex number which does not belong to the spectrum of QHQ. Then it is possible to show that $z \in \text{Sp}(H) - \text{Sp}(QHQ)$ if and only if $z \in \text{Sp}(H_{\text{eff}}(z))$. Moreover E is an eigenvalue of H not in $\text{Sp}(QHQ)$ if and only if E is an eigenvalue of $H_{\text{eff}}(E)$.

We then denote by $P = I - Q$ the projection $I \otimes P_j(0)$ of $\mathcal{A}_I(0) \otimes M_q$. For $(\mathbf{k}; \gamma) \approx (0, 0)$, it follows that there is a small neighbourhood O of $e_j(0)$ such that if $z \in O$, $z \notin \text{Sp}(QH_{\text{scl}}(\mathbf{k}; \gamma)Q)$. Since the eigenvalue $e_j(\mathbf{k})$ is simple for $\mathbf{k} \approx 0$, the projector $P_j(\mathbf{k})$ is one dimensional for $\mathbf{k} \approx 0$, and therefore there exists a partial isometry $\hat{D} : \mathbf{C}^q \mapsto \mathbf{C}$ such that $\hat{D}\hat{D}^* = I$, and $\hat{D}^*\hat{D} = P_j(0)$. If D is the partial isometry $I \otimes \hat{D}$, let us introduce the effective Hamiltonian :

$$h_j(z) = DH_\gamma D^* + DH_\gamma Q (zI - QH_\gamma Q)^{-1} QH_\gamma D^* . \quad (117)$$

By construction this is an analytic family of elements in $\mathcal{A}_I(0)$ now. We can therefore analyze it by the method developed in 5.2, and will give rise to a bundle of Landau sublevels $E_{n,j}(z)$ near the lower edge of the band B_j. The corresponding part of the spectrum of H_γ near $e_j(0) = 0$, will then be given by solving the implicit equation $E = E_{n,j}(E)$. The solution can be computed explicitly order by order in powers of γ, thanks to the hypothesis made on e_j. The Rammal-Wilkinson term comes from the first order contribution of the second term in (117). It reflects the fact that the matrices $H_{\text{scl}}(\mathbf{k}; \gamma)$ do not mutually commute for various values of $\mathbf{k}$ in general, namely it reflects the existence of a curvature in the fiber bundle over the 2-torus defined by $P_j(\mathbf{k})$. The calculation of this term can be found in [RaBe:Alg, BeKrSe].

5.4 Qualitative analysis of the spectrum

Let us now comment on the formulæ (114)&(115). Due to the absolute value of γ appearing in the first term of (114), the right and left derivatives of the band edge with respect to γ are different, showing that the band edges eventhough continuous functions of γ by the theorem 3, have nevertheless a discontinuous first derivative at each rational point. On the other hand, even if $\partial_\gamma H = 0$ the Rammal-Wilkinson term may not vanish. This is the case for instance in the Harper model for $p/q = 1/3$ (see Fig. 1). We can see the effect of this term by the fact that the left and right derivative of the band edge are not symmetric around $\gamma = 2\pi p/q$. The difference between them reveals the occurrence of curvature effects.

On the other hand one can recognize whether the band edge is a maximum or a mini at the slope of the Landau sublevels emerging away from $\gamma = 2\pi p/q$.

For most values of p/q, all bands are separated by gaps. However, many non generic situation can be observed on examples.

(i) Two bands may overlap without touching each other. Then, each minimum or each maximum of the corresponding band will reveal itself by the occurence of a bundle of Landau levels emerging on both sides of $\gamma = 2\pi p/q$ (see Fig. 5), and given by the formula (114)&(115).

(ii) two bands B_j, $B_{j'}$, with or without overlap, may touch each other. In this case, generically they will touch on a conical point (see Fig. 6). This situation leads to a different canonical form. For indeed the previous analysis can be extended by replacing the projector $P_j(0)$ by $P_j(0) + P_{j'}(0)$. Then the effective Hamiltonian becomes a 2×2 matrix unitarily equivalent to the Dirac operator [HeSj:Har2, RaBe:Alg] :

$$H_{\text{Dirac}} = |\gamma|^{1/2} \begin{bmatrix} 0 & K_1 + iK_2 \\ K_1 - iK_2 & 0 \end{bmatrix} + O(\gamma) . \tag{118}$$

This case will give "Dirac levels" which are parabolic namely :

$$E_{\pm n} = \pm \text{const.} |n\gamma|^{1/2} , \, n \in \mathbf{N} , \tag{119}$$

which is for instance what happens in the Harper model at $E = 0$ and $p/q = 1/2$ (see Fig. 1).

This formula must usually be corrected by a Rammal-Wilkinson term, giving a slope to the sublevel $n = 0$. This is what happens in the WBK-model (108), at $p/q = 1/2$ (see Fig. 7).

(iii) Two bands can also touch with a contact of order 2. There is another example proposed by M.Wilkinson [Wil:Cri] and studied in details by Barelli and Fleckinger [BaFl], which is the following :

$$H_{\text{W}} = T_1 + T_2 + t_3 \left(T_1^2 T_2 e^{-i\gamma} + T_1^{-2} T_2 e^{i\gamma} + T_1 T_2^2 e^{-i\gamma} + T_1 T_2^{-2} e^{i\gamma} \right) + \text{h.c.} . \tag{120}$$

At $E = 0, p/q = 1/2$, we do get two families of Landau sublevels on either side of $p/q = 1/2$, corresponding to the bottom wells of the two bands. The generic parabolic touching can be seen on Fig. 8.

(iv) A maximum or a minimum can also be reached on a curve. This has been observed in the WBK model at $p/q = 1/2$. This case has been investigated in details by Helffer and Sjöstrand [HeSj:Har3], who remarked that the "subprincipal symbol" may break this degeneracy and create what they have called "miniwells", namely local extremas with deepness of order $O(\gamma)$. Such an example has never been investigated numerically, but there are indications that such a phenomenon should occur on the WBK model.

At last we must point out the occurrence of tunneling effect. For indeed, the classical model gives a Hamiltonian on the phase space given by a 2-torus. This is equivalent to choosing $\mathbf{R}^2$ instead, but requiring that the Hamiltonian be periodic in both directions. This will be called the "extended picture". In this picture, each local extremum is repeated periodically, giving rise to an exact degeneracy. Therefore a tunneling effect should occur between the corresponding wells, ending into a broadening of the Landau levels or sublevels. The width of this broadening can be computed by the WKB method, and will give rise to terms of order $O(\exp(-S/\gamma))$ where S is some constant equal to the real part of the tunneling action between two neighbouring wells.

This effect has been studied in great details in the Harper model by Helffer & Sjöstrand [HeSj:Har1, HeSj:Har2, HeSj:Har3], and for the corresponding model on a honeycomb or triangular lattice by Kerdelhué [Ker]. By evaluating precisely the tunneling matrix representing the effective Hamiltonian restricted to each of the Landau sublevel, they could prove that it is again represented by a Hamiltonian with nearest neighbour interactions, having the symmetry of the original lattice (e.g. a Harper model for a square lattice), with a small correction. Therefore, each Landau sublevel is itself decomposed into subbands, and this explain the occurence of the fractal structure.

This tunneling effect has also been exhibited in a spectacular example by Barelli & Kreft [BaKr], in the WBK model for $t_2 > 1/4$ and $\gamma \approx 0$. As we already said, after the bifurcation the unique minimum splits into four degenerate minima surrounding one maximum. Since these four wells are very close to each other in each unit cell of the extended phase space, compare to the distance between cells, the tunneling effect between these four wells within the unit cell is likely to dominate over the other sources of tunneling. Each well gives rise to its own bunch of Landau levels, but the splitting due to the tunneling will separate them. It turns out that the tunneling action in this case is not purely imaginary, so that the Landau levels can be represented by if $n \in \mathbf{N}$, $t_2 = 1/2$ and $i = 1, 2, 3, 4$:

$$E_{n,i} = E_n(\gamma) + dE_{n,i}(\gamma) \ , \ E_n(\gamma) = -3 + \frac{3}{2}\gamma(2n+1) + O(\gamma^2) \ , \tag{121}$$

where the splitting is given by [BaKr] :

$$dE_{n,i} = \gamma\frac{3}{\pi}e^{-\operatorname{Im}(S_2)/\gamma}\cos(\operatorname{Re}(S_4)/4\gamma + \pi/4) + O(e^{-S'/\gamma}) \ , \tag{122}$$

where S_2 represents the action $\int_{AB} k_1 dk_2$ for a path AB in the complex energy surface $H_d(\mathbf{k}) = E_n(\gamma)$ joining two neighbouring wells A and B, while S_4 is the tunneling action for a closed path in the same energy surface going through the four wells once. Moreover, S' is some action larger than S_2. Even though there are usually many non homotopic such paths in this complex energy surface, only the "shortest" ones (in terms of the corresponding action integral) do contribute to this order.

In this formula the width of the splitting is controlled by $\operatorname{Im}(S_2)$ which gives an exponentially small term. But the occurence of a non zero real part produces a nice braiding between these four sublevels as can be seen in (Fig. 9). In a recent work, Barelli and Fleckinger exhibited a braiding of Dirac sublevels near the half flux (see Fig. 10) [BaFl].

6 Elementary Properties of the Kicked Rotor

6.1 The Furstenberg Algebra

As we have seen the Floquet operator for the kicked rotor cannot be seen as an element of the rotation algebra. This is because the kinetic part is not a continuous function of U and V. However, we have seen that it defines a *-automorphism of the rotation algebra. To deal with that we have two choices. The first is to ignore the Floquet operator itself and to stick with its action on the non commutative torus. This is fine as long as we are interested only in the evolution of observables.

However, in many occasions do physicists need to know more on the spectrum of the Floquet operator itself, the so-called "quasi-energy" spectrum. One of its most important property is the "dynamical localization", a phenomenon similar to the Anderson localization in Solid State Physics of disordered metals [FiGrPr].

In order to deal with this latter problem, we can simply enlarge our algebra by brute force, adding the missing unitary F_0 equal to the kinetic energy defined in section 1 (12) by :

$$F_0 = e^{-iA^2/2\gamma} . \tag{123}$$

As we have seen in section 1 (19) this operator satisfies the following commutation rules

$$\text{(i)} \quad F_0 V F_0^{-1} = V \qquad \text{(ii)} \quad F_0 U F_0^{-1} = UV^{-1} e^{-i\gamma/2} . \tag{124}$$

As before we will denote by $\mathcal{B}_I$ the C^*-algebra generated by the polynomials in U, V, F_0 with coefficients in the set of continuous functions of γ in I. This algebra can be rigorously constructed along the line developed in section 2. However one can use the general method of C^*-algebras, namely the notion of crossed-product [Ped], to construct it.

One can indeed see $\mathcal{B}_I$ in two ways :
(i)-the first one comes from the previous definition, namely F_0 acts on the rotation algebra $\mathcal{A}_I$ by mean of the *-automorphism

$$\beta_0(a) = F_0 a F_0^{-1} \qquad a \in \mathcal{A} . \tag{125}$$

Therefore $\mathcal{B}_I$ can be seen as the crossed product $\mathcal{A}_I \times_{\beta_0} \mathbf{Z}$ of the rotation algebra $\mathcal{A}_I$ by the $\mathbf{Z}$-action defined by β_0. Using Weyl's operators defined in section 2 (32), we notice that

$$\beta_0 (W(\mathbf{m})) = W(G\mathbf{m}) , \qquad \mathbf{m} \in \mathbf{Z}^2 , \tag{126}$$

provided G is the element of $SL(2, \mathbf{R})$ given by :

$$G = \begin{bmatrix} 1 & 0 \\ -1 & 1 \end{bmatrix} . \tag{127}$$

(ii)-the second one consists in considering first the subalgebra generated by functions in $\mathcal{C}(I)$, together with the operators V and F_0. This is an abelian C^*-algebra isomorphic to $\mathcal{C}(I \times \mathbf{T}^2)$. This isomorphism associates to V and F_0 respectively the functions $f_V(\gamma, x, y) = e^{ix}$ and $f_{F_0}(\gamma, x, y) = e^{iy}$. Actually, the inner automorphism associated to U leaves this algebra invariant. This is because the commutation rules (19) can be written as

$$\text{(i)} \quad UVU^{-1} = e^{i\gamma}V , \qquad \text{(ii)} \quad UF_0U^{-1} = e^{i\gamma/2}VF_0 . \tag{128}$$

In other words, for $f \in \mathcal{C}(I \times \mathbf{T}^2)$, we get :

$$UfU^{-1} = f \circ \phi , \tag{129}$$

where ϕ is the "Furstenberg" map acting on $I \times \mathbf{T}^2$ as :

$$\phi(\gamma, x, y) = (\gamma, x + \gamma, y + x + \gamma/2) , \qquad (\gamma, x, y) \in I \times \mathbf{T}^2 . \tag{130}$$

This map was used by Furstenberg to study the ergodic properties of diophantine approximations in number theory. Thus $\mathcal{B}_I$ can be seen as the crossed product $\mathcal{C}(I \times \mathbf{T}^2) \times_\phi \mathbf{Z}$ by the Furstenberg map. This is why we propose to call this algebra the "Furstenberg algebra".

We see that ϕ leaves each fiber $\{\gamma\} \times \mathbf{T}^2$ invariant and we will denote by ϕ_γ the corresponding restriction. It is well-known that whenever $\gamma/2\pi$ is irrational, ϕ_γ is a minimal diffeomorphism [CoFoSi].

6.2 Calculus on $\mathcal{B}_I$

As for $\mathcal{A}_I$, a calculus can be defined on the Furstenberg algebra. Since the trace on $\mathcal{A}_I$ is β_0-invariant, it defines a trace on the crossed product in a natural way. It is actually defined by the formula :

$$\tau\left(W(\mathbf{m})F_0^l\right) = \delta_{\mathbf{m},0}.\delta_{l,0} , \qquad \mathbf{m} \in \mathbf{Z}^2 , l \in \mathbf{Z} . \tag{131}$$

Since we have defined originally (cf. section 1) U, V, F_0 in term of action-angle variables in the classical case, one also gets an angle average $\langle \cdot \rangle$ namely :

$$\langle W(\mathbf{m})F_0^l \rangle = \delta_{m_1,0} V^{m_2} F_0^l , \qquad if \quad \mathbf{m} = (m_1, m_2) \in \mathbf{Z}^2 , l \in \mathbf{Z} . \tag{132}$$

Thus, if $a \in \mathcal{B}_I$, $\langle a \rangle \in \mathcal{C}(I \times \mathbf{T}^2)$, and this average satisfies the properties described in (32).

In much the same way, a differential structure can be defined. The derivation ∂_θ can be extended immediately to $\mathcal{B}_I$ by :

$$\partial_\theta U = iU , \qquad \partial_\theta V = 0 , \qquad \partial_\theta F_0 = 0 . \tag{133}$$

We notice however that ∂_A cannot be extended as a derivation in $\mathcal{B}_I$ because $\partial_A F_0$ would be unbounded, namely outside $\mathcal{B}_I$. But a new derivation ∂_y appears defined by :

$$\partial_y U = 0 , \qquad \partial_y V = 0 , \qquad \partial_y F_0 = iF_0 . \tag{134}$$

Both ∂_θ and ∂_y are the infinitesimal generators of the following two-parameter group of $*$-automorphisms :

$$\hat{\rho}_{\theta,y}\left(W(\mathbf{m})F_0^l\right) = e^{i(m_1\theta + ly)} W(\mathbf{m})F_0^l , \tag{135}$$

which leaves the trace invariant.

At last, the definition of a Poisson bracket is not obvious because for $\gamma = 0$ the algebra $\mathcal{B}_0$ is no longer commutative. Even though it is in principle possible to define such an object, we will not use it, and we skip this part of the calculus.

6.3   Representations and structure of $\mathcal{B}_I$

Among the representations of $\mathcal{B}_I$, we will select one family of special interest in view of the original definition of the kicked rotor in the physical Hilbert space $L^2(\mathbf{T})$ given in section 1. It is actually simpler to work in the momentum space, namely in $\ell^2(\mathbf{Z})$ where the integers of the chain $\mathbf{Z}$ are simply the quantum numbers for the angular momentum.

This family $\{\pi_{\gamma,x,y}; (\gamma, x, y) \in I \times \mathbf{T}^2\}$ is indexed by points in $I \times \mathbf{T}^2$ and acts on $\ell^2(\mathbf{Z})$ as follows :

$$
\begin{aligned}
&\text{(i)} && (\pi_{\gamma,\mathrm{x},\mathrm{y}}(\mathrm{f})\psi)\,(\mathrm{n}) = \mathrm{f}(\gamma)\psi(\mathrm{n}) \,, && \mathrm{f} \in \mathcal{C}(I) \\
&\text{(ii)} && (\pi_{\gamma,\mathrm{x},\mathrm{y}}(\mathrm{U})\psi)\,(\mathrm{n}) = \psi(\mathrm{n} - 1) \,, \\
&\text{(iii)} && (\pi_{\gamma,\mathrm{x},\mathrm{y}}(\mathrm{V})\psi)\,(\mathrm{n}) = \mathrm{e}^{\mathrm{i}(\mathrm{x}-\mathrm{n}\gamma)}\psi(\mathrm{n}) \,, \\
&\text{(iv)} && (\pi_{\gamma,\mathrm{x},\mathrm{y}}(\mathrm{F_0})\psi)\,(\mathrm{n}) = \mathrm{e}^{\mathrm{i}(\mathrm{y}-\mathrm{nx}+\mathrm{n}^2\gamma/2)}\psi(\mathrm{n}) \,, && if \quad \psi \in \ell^2(\mathbf{Z}) \,.
\end{aligned}
\tag{136}
$$

Comparing with the equation (12) & (13), γ appears as an effective Planck constant, x as an effective magnetic field, and y as a phase factor entering in the definition of F_0.
With these definitions, the following result can be easily proved by standard technics :

Proposition 7 *1)-The family $\{\pi_{\gamma,x,y}; (\gamma, x, y) \in I \times \mathbf{T}^2\}$ is faithfull. In particular, the norm of $a \in \mathcal{B}_I$ is given by :*

$$
\|a\| = \sup_{\gamma \in I} \sup_{(x,y) \in \mathbf{T}^2} \|\pi_{\gamma,x,y}(a)\| \,.
\tag{137}
$$

2)-The map $(\gamma, x, y) \in I \times \mathbf{T}^2 \mapsto \pi_{\gamma,x,y}(a)$ is strongly continuous for all $a \in \mathcal{B}_I$.
3)-For $\gamma \in I$, the trace is given by :

$$
\tau_\gamma(a) = \int_{\mathbf{T}^2} \frac{dxdy}{4\pi^2} \ \langle 0|\pi_{\gamma,x,y}(a)|0\rangle
\tag{138}
$$

Moreover if $\gamma/2\pi$ is irrational, we get

$$
\tau_\gamma(a) = \lim_{L \mapsto \infty} \frac{1}{2L + 1} \mathrm{Tr}\left(\pi_{\gamma,x,y}(a)|_{[-L,L]}\right) \,,
$$

uniformly in $(x, y) \in \mathbf{T}^2$. 4)-If T is the translation operator in $\ell^2(\mathbf{Z})$, namely if $(T\psi)(n) = \psi(n - 1)$ for $\psi \in \ell^2(\mathbf{Z})$, then :

$$
T\pi_{\gamma,x,y}(a)T^{-1} = \pi_{\phi(\gamma,x,y)}(a) \,, \qquad a \in \mathcal{B}_I \,, \ (\gamma, x, y) \in I \times \mathbf{T}^2 \,.
\tag{139}
$$

5)-If N is the position operator in $\ell^2(\mathbf{Z})$ defined by $(N\psi)(n) = n\psi(n)$, $\psi \in \ell^2(\mathbf{Z})$, we have :

$$
\pi_{\gamma,x,y}(\partial_\theta a) = i[N, \pi_{\gamma,x,y}(a)] \,, \qquad \pi_{\gamma,x,y}(\partial_y a) = \frac{\partial}{\partial y}\pi_{\gamma,x,y}(a) \,.
\tag{140}
$$

Thanks to this result the elements of $\mathcal{B}_I$ can be described as follows. For $a \in \mathcal{B}_I$, we set :

$$
a(\gamma, x, y; n) = \langle 0|\pi_{\gamma,x,y}(a)|n\rangle \,.
\tag{141}
$$

This is a continuous function on $I \times \mathbf{T}^2 \times \mathbf{Z}$ converging to zero at infinity. In terms of such functions the product and the $*$ in $\mathcal{B}_I$ can be expressed as follows :

$$ab(\gamma, x, y; n) = \sum_{l \in \mathbf{Z}} a(\gamma, x, y; l)b(\gamma, x - l\gamma, y - lx + l^2\gamma/2; n - l) , \qquad (142)$$

$$a^*(\gamma, x, y; n) = a(\gamma, x - n\gamma, y - nx + n^2\gamma/2; -n)^* , \qquad (143)$$

for $a, b \in \mathcal{B}_I$. Moreover, the representation $\pi_{\gamma, x, y}$ is given by :

$$(\pi_{\gamma, x, y}(a)\psi)(n) = \sum_{l \in \mathbf{Z}} a(\gamma, x - n\gamma, y - nx + n^2\gamma/2; l - n)\psi(l) , \qquad \psi \in \ell^2(\mathbf{Z}) . \quad (144)$$

In particular, due to the faithfullness of this family, $a = 0$ if and only if the function $a(\gamma, x, y; n)$ vanishes identically.

If we denote by $\mathcal{B}_\gamma$ the algebra $\mathcal{B}_I$ for $I = \{\gamma\}$, the following theorem characterizes its structure :

Theorem 11 *1)-If $\gamma/2\pi$ is irrational, $\mathcal{B}_\gamma$ is simple. In particular, every non zero representation is faithfull.*
2)-For $\gamma = 0$, the algebra $\mathcal{B}_0$ is isomorphic to the universal rotation algebra $\mathcal{A}$.
3)-If $\gamma = 2\pi p/q$ where p, q are positive integers prime to each others, $\mathcal{B}_{2\pi p/q}$ is isomorphic to the sub C^*-algebra of $M_q(\mathbf{C}) \otimes \mathcal{B}_0$ generated by :

$$\tilde{U} = u \otimes U_0 , \qquad \tilde{V} = v \otimes V_0 , \qquad \tilde{F}_0 = w \otimes F_{0,0} , \quad (145)$$

where $U_\gamma, V_\gamma, F_{0,\gamma}$ are the generators of $\mathcal{B}_\gamma$, and u, v, w are three unitary $q \times q$ matrices fulfilling the following conditions :

$$u^q = v^q = w^{2q} = \mathbf{I} , \qquad (146)$$

$$uvu^{-1} = e^{2i\pi p/q}v , \qquad uwu^{-1} = e^{i\pi p/q}vw , \qquad vw = wv . \quad (147)$$

Proof : 1)-For $\gamma/2\pi$ irrational, the Furstenberg map $\phi_\gamma : (x, y) \in \mathbf{T}^2 \mapsto (x + \gamma, y + x + \gamma/2) \in T^2$ is a minimal diffeomorphism of the torus [CoFoSi]. Therefore, the crossed product $\mathcal{B}_\gamma = C(\mathbf{T}^2) \times_{\phi_\gamma} \mathbf{Z}$ is simple [HiSk].
2)-For $\gamma = 0$ the commutation rules become :

$$UV = VU , \qquad VF_0 = F_0V , \qquad UF_0U^{-1} = VF_0 . \quad (148)$$

These rules are precisely the ones defining the universal rotation algebra $\mathcal{A}$ if we identify V with the map $\gamma \in \mathbf{T} \mapsto e^{i\gamma} \in \mathbf{C}$ (cf. section 2).
3)-If one chooses the matrices u, v as in (84), the matrix w becomes :

$$w = \begin{bmatrix} 1 & 0 & 0 & \cdots & 0 & 0 \\ 0 & \lambda' & 0 & \cdots & 0 & 0 \\ \cdots & \cdots & \cdots & \cdots & \cdots & \cdots \\ \cdots & \cdots & \cdots & \cdots & \cdots & \cdots \\ 0 & 0 & 0 & \cdots & \lambda'^{(q-2)^2} & 0 \\ 0 & 0 & 0 & \cdots & 0 & \lambda'^{(q-1)^2} \end{bmatrix} , \quad (149)$$

where $\lambda' = e^{i\pi p/q}$.

It is easy to check that $\tilde{U}, \tilde{V}, \tilde{F}_0$ satisfy the commutation rules for the algebra $\mathcal{B}_{2\pi p/q}$. Hence they define a $*$-homomorphism ρ from $\mathcal{B}_{2\pi p/q}$ into $M_q(\mathbf{C}) \otimes \mathcal{B}_0$.
4)-To achieve our result it is sufficient to prove that ρ is one-to-one. For $(x,y) \in \mathbf{T}^2$, let $\hat{\pi}_{x,y}$ be the representation of $M_q(\mathbf{C}) \otimes \mathcal{B}_0$ given by $id \otimes \pi_{0,x,y}$ acting on $\mathbf{C}^q \otimes \ell^2(\mathbf{Z})$. Any $a \in \mathcal{B}_0$ can be seen as a function on $\mathbf{T}^2 \times \mathbf{Z}$ as (see (141)), and for $\phi \in \mathbf{C}^q \otimes \ell^2(\mathbf{Z})$ and $A \in M_q$ we get :

$$[\hat{\pi}_{x,y}(A \otimes a)\phi]_j(n) = \sum_{j=0}^{q-1} \sum_{l\in\mathbf{Z}} A_{j,j'}a(x, y - nx; l - n)\phi_{j'}(l) . \tag{150}$$

Let $\{e_j\}_{j=0}^{q-1}$ be the canonical basis of $\mathbf{C}^q$ with the convention that $e_{j+q} = e_j$, and let $\{\delta_n; n \in \mathbf{Z}\}$ be the canonical basis of $\ell^2(\mathbf{Z})$. We set :

$$|j,n\rangle = e_j \otimes \delta_n . \tag{151}$$

Then :

$$\langle j,0|\hat{\pi}_{x,y}(A \otimes a)|j',l\rangle = A_{j,j'}a(x,y,;l) . \tag{152}$$

It is not difficult to check that if now $b \in \mathcal{B}_{2\pi p/q}$ and $\phi \in \mathbf{C}^q \otimes \ell^2(\mathbf{Z})$ we get :

$$[\hat{\pi}_{x,y}(\rho(b))\phi]_j(n) = \sum_{l\in\mathbf{Z}} b(x - 2\pi jp/q, y - nx + j^2\pi p/q; l)\phi_{j+l}(n+l) , \tag{153}$$

where $j + l$ is defined modulo q. It is actually sufficient to check this formula on the generators $U_{2\pi p/q}, V_{2\pi p/q}, F_{0,2\pi p/q}$ since $\hat{\pi}_{x,y}$ and ρ are $*$-homomorphisms. In particular :

$$\langle 0,0|\hat{\pi}_{x,y}(\rho(b))|l,l\rangle = b(x,y;l) . \tag{154}$$

Thus $\rho(b) = 0$ if and only if $b(x,y;l) = 0$ for any $(x,y;l)$, namely $b = 0$. Hence ρ is one-to-one.

Using the same strategy we can easily get :

Corollary 4 *1)-for $\gamma \in \mathbf{R}$ the algebra $\mathcal{B}_{2\pi p/q+\gamma}$ is isomorphic to the subalgebra of $M_q \otimes \mathcal{B}_\gamma$ generated by $u \otimes U_\gamma, v \otimes V_\gamma, w \otimes F_{0,\gamma}$.*
2)-for $\gamma, \gamma' \in \mathbf{R}$ the algebra $\mathcal{B}_{\gamma+\gamma'}$ is isomorphic to the subalgebra of $\mathcal{B}_\gamma \otimes \mathcal{B}_{\gamma'}$ generated by $U_\gamma \otimes U_{\gamma'}, V_\gamma \otimes V_{\gamma'}, F_{0,\gamma} \otimes F_{0,\gamma'}$.

6.4 Algebraic Properties of the Kicked Rotor

In section 1 we have expressed the Floquet operator of the kicked rotor as :

$$F_{K,\gamma,x}^{-1} = e^{-iA^2/2\gamma} e^{-iK\cos\theta/\gamma} e^{i\hat{y}} , \tag{155}$$

where $\gamma = \hbar T/I$ is the effective Planck constant and $x = -\mu BT$ is the effective magnetic field. Moreover in the momentum space representation, $A = \gamma N - x$ if N is the position operator (see prop. 7). Using the previous algebraic framework, it follows that :

$$F_{K,\gamma,x} = \pi_{\gamma,x,0}(F_K) , \tag{156}$$

with :

$$F_K = e^{iK(U+U^{-1})/2\gamma} F_0 , \tag{157}$$

where $\hat{\gamma} : \gamma \in I \mapsto \gamma \in \mathbf{R}$. In this special case we notice the following property :

$$\pi_{\gamma,x,y}(F_K) = e^{iy}F_{K,\gamma,x} , \tag{158}$$

so that one can set $y = 0$ without loss of generality.

It follows that F_K belongs to $\mathcal{B}_I$ for any compact set I in the real line not containing the origin.

Our first set of results concerns the spectrum as a set of this Floquet operator. Since it is unitary its spectrum is necessarily contained in the unit circle S_1. Actually the following results are still valid if we replace $cos(\theta) = (U + U_{-1})/2$ by any real valued 2π-periodic continuous function $g(\theta)$ on the real line.

Theorem 12 *1)-For any $\gamma \neq 0$, the spectrum of $F_{K,\gamma} = \eta_\gamma(F_K)$ is the full circle. 2)-If $\gamma/2\pi$ is irrational, the spectrum of $F_{K,\gamma,x}$ is the full circle for any $x \in \mathbf{T}$. 3)-If $\gamma/2\pi$ is rational, but $x/2\pi$ is irrational, the spectrum of $F_{K,\gamma,x}$ is the full circle. 4)-If $\gamma/2\pi$ and $x/2\pi$ are rational, $F_{K,\gamma,x}$ admits a band spectrum.*

Proof : 1)-Since the family $\{\pi_{\gamma,x,y}; (x,y) \in \mathbf{T}^2\}$ of representations of $\mathcal{B}_\gamma$ is faithfull, we get :

$$\mathrm{Sp}_{\mathcal{B}_\gamma}(F_{K,\gamma}) = \cup_{(x,y)\in\mathbf{T}^2} e^{iy}\mathrm{Sp}(F_{K,\gamma,x}) . \tag{159}$$

Taking the union over y clearly gives the full circle.

2)-If $\gamma/2\pi$ is irrational, $\mathcal{B}_\gamma$ is simple. Thus each of the $\pi_{\gamma,x,y}$'s is faithfull, in particular :

$$S_1 = \mathrm{Sp}_{\mathcal{B}_\gamma}(F_{K,\gamma}) = \mathrm{Sp}(F_{K,\gamma,x}) , \qquad \forall x \in \mathbf{T}^2 . \tag{160}$$

3)-If $\gamma = 2\pi p/q$, the covariance condition (145) gives :

$$T^{nq}\pi_{\gamma,x,y}(F_K)T^{-nq} = \pi_{\gamma,x,y+nqx}(F_K) , \qquad n \in \mathbf{Z} . \tag{161}$$

In particular,

$$\mathrm{Sp}\,(\pi_{\gamma,x,y}(F_K)) = \mathrm{Sp}\,(\pi_{\gamma,x,y+nqx}(F_K)) , \qquad \forall n \in \mathbf{Z} . \tag{162}$$

If in addition $x/2\pi$ is irrational, given any $y' \in \mathbf{T}$ we can find a sequence (n_l) of integers such that $y' - y = \lim_{l\to\infty} n_l qx \mod 2\pi$. By the strong continuity of $\pi_{\gamma,x,y}$ with respect to y, it follows that :

$$\mathrm{Sp}\,(\pi_{\gamma,x,y'}(F_K)) \subset \mathrm{Sp}\,(\pi_{\gamma,x,y}(F_K)) . \tag{163}$$

Since y, y' are arbitrary, the same result holds after exchanging them. In particular for any y we have :

$$\mathrm{Sp}\,(\pi_{\gamma,x,y}(F_K)) = e^{iy}\mathrm{Sp}\,(\pi_{\gamma,x,0}(F_K)) = \mathrm{Sp}\,(\pi_{\gamma,x,0}(F_K)) , \tag{164}$$

showing the result.

4)-If $\gamma = 2\pi p/q$ and $x = 2\pi r/s$, the covariance property shows that $\pi_{\gamma,x,0}(F_K)$ is periodic. By the Bloch theorem we get a band spectrum. Actually one can easily see, using the corollary 4 that the algebra $\pi_{\gamma,x,0}(\mathcal{B}_I)$ is isomorphic to the subalgbera of $M_q \otimes M_s \otimes \mathcal{C}(\mathbf{T})$ generated by $u \otimes u' \otimes e^{ik}, v \otimes e^{ix} \otimes 1, w \otimes v' \otimes 1$ where u, v, w (resp. u', v') are the $q \times q$ matrices (resp. $s \times s$) defined in the theorem 12, and k is the quasimomentum. Here we used the fact that $\pi_{\gamma,x,0}(F_K)^Q = \mathbf{I}$ if $Q = 2(q \vee s)$. This gives the band spectrum by diagonalizing the finite dimensional matrices and varying k.

The next set of results concerns the density of states. Let Δ be an interval in the unit circle, namely the image by $\omega \mapsto e^{i\omega}$ of an interval of the real line. Let us also call g_L the restriction to the finite set $[-L, L]$ of $g(\theta)$. This is a self adjoint matrix of dimension $2L + 1$. Let also $F_{0,\gamma,x}^{(L)}$ be the restriction of $F_{0,\gamma,x}$ to the same interval. Because it is diagonal it is a unitary $(2L+1) \times (2L+1)$ matrix. Then we set $F_{K,\gamma,x}^{(L)} = F_{0,\gamma,x}^{(L)} e^{iKg_L(\theta)/\gamma}$. Again, this is a unitary matrix of dimension $2L+1$. Let then $n_L(\Delta)$ be the number of eigenvalues of this matrix contained in Δ. As $L \mapsto \infty$, this number increases like $O(L)$, so that we can define the Integrated Density of States (IDS) as the following limit, if it exists :

$$\mathcal{N}_{\gamma,x,y}(\Delta) = \lim_{L \mapsto \infty} \frac{n_L(\Delta)}{2L + 1} . \qquad (165)$$

The first important property is the "Shubin formula" [Bel:Gap]

Proposition 8 *If $\gamma/2\pi$ is irrational, the limit defining the IDS exists uniformly with respect to $(x,y) \in \mathbf{T}^2$ and is independent of (x,y). Moreover it is equal to :*

$$\mathcal{N}_{\gamma}(\Delta) = \tau_{\gamma}\left(\chi_{\Delta}(F_K)\right) , \qquad \text{(Shubin's Formula)} . \qquad (166)$$

where χ_{Δ} is the characteristic function of the interval Δ.

The proof of this proposition can be found in [Bel:Kth, Bel:Gap, BeBoGh] for self adjoint operators. It can be easily adapt for the Floquet operator. We notice that the limit is reached uniformly with respect to (x,y). This is because the Furstenberg map is minimal and not only ergodic. Another remark is that the eigenprojection $\chi_{\Delta}(F_K)$ does not belong in general to the algebra $\mathcal{B}_{\gamma}$. However, it belongs to the von Neumann algebra $L^{\infty}(\mathcal{B}_{\gamma}, \tau_{\gamma})$, namely the weak closure of $\mathcal{B}_{\gamma}$ in the GNS representation associated to the trace. Thus the Shubin formula is meaningfull.

Thanks to the Shubin formula, the IDS can be written as :

$$\mathcal{N}_{\gamma}(\Delta) = \int_{\Delta} d\mathcal{N}_{\gamma}(E) , \qquad (167)$$

where $d\mathcal{N}_{\gamma}$ is a probability measure on the torus $\mathbf{T}$ (which we identify with the unit circle) called the Density of States (DOS). We can actually compute the DOS namely :

Proposition 9 *If $\gamma/2\pi$ is irrational, for any continuous real valued 2π-periodic function g on the real line, the DOS of the kicked rotor is equal to the normalized Lebesgue measure on the torus, namely :*

$$d\mathcal{N}_{\gamma}(E) = \frac{dE}{2\pi} . \qquad (168)$$

Proof : The Shubin formula implies that the DOS is the unique probability measure on the torus such that :

$$\int_{\mathbf{T}} d\mathcal{N}_\gamma(E) e^{inE} = \tau_\gamma(F_K^n) , \qquad n \in \mathbf{Z} . \tag{169}$$

We claim that $\tau_\gamma(F_K^n) = 0$ unless $n = 0$ which will prove the result. For indeed, the trace is invariant by the automorphism group $\hat{\rho}_{0,k}$. On the other hand, we have :

$$\hat{\rho}_{0,k}(F_K) = e^{ik} F_K . \tag{170}$$

It follows that $\hat{\rho}_{0,k}(F_K^n) = e^{ink} F_K^n$ showing that

$$\tau_\gamma(F_K^n)(e^{ink} - 1) = 0 . \tag{171}$$

Our last result concerns the algebraic way of writing the kinetic energy. In order to study numerically the spectral properties of the kicked rotor, several physicists [CaChIzFo] have introduced the averaged kinetic energy. Giving an initial state $\phi \in \ell^2(\mathbf{Z})$, it is given by (see (9) & (11)) :

$$\mathcal{E}_c(t) = \langle\phi|F^t \frac{\mathbf{L}^2}{2I} F^{-t}|\phi\rangle , \qquad t \in \mathbf{Z} , \tag{172}$$

where $\mathbf{L}$ is the angular momentum, I is the moment of inertia and F the Floquet operator. Thanks to the definition of the position operator N (see Prop.7) and introducing the period T of the kicks, one can write it as :

$$\mathcal{E}_c(t) = \frac{I}{2T^2} \langle\phi|F^t \gamma^2 N^2 F^{-t}|\phi\rangle . \tag{173}$$

In order to keep only dimensionless quantities, we will redefine this kinetic energy by forgetting the prefactor $I/2T^2$. Moreover physicists usually choose an initial state localized on one value of the initial angular momentum. Using the covariance condition, it is always possible to choose $\phi = |0\rangle$ by changing the value of (x, y) if necessary. This why we will rather define the mean kinetic energy in the following way :

$$\mathcal{E}_{\gamma,x}(t) = \gamma^2 \langle 0|F_{K,\gamma,x}^t N^2 F_{K,\gamma,x}^{-t}|0\rangle . \tag{174}$$

We notice that varying y will not change this definition. Using now (146), it follows immediately that if $|A|^2 = AA^*$:

$$\mathcal{E}_{\gamma,x}(t) = \gamma^2 \langle 0|\pi_{\gamma,x,y}(|\partial_\theta F_{K,\gamma,x}^t|^2|0\rangle . \tag{175}$$

The choice of the initial value of the angular momentum being arbitrary, we may average over the position of the initial state in momentum space, in order to get the generic properties of the system. This is equivalent to averaging over (x, y), namely to taking the trace. This why we will also consider the quantity :

$$\mathcal{E}_\gamma(t) = \gamma^2 \tau_\gamma\left(|\partial_\theta F_{K,\gamma,x}^t|^2\right) . \tag{176}$$

7 Localization and Dynamical Localization

7.1 Anderson's Localization

The localization phenomena was predicted in 1958 by Anderson [And] for conduction electrons in a disordered metal. The main idea underlying this effect is that the electronic wave in an infinite medium is reflected by the obstacles (ions, defects,etc,...). If the medium is a perfect crystal, the total reflection coefficient may not be equal to one due to constructive interference effects and allows the wave to travel freely towards the boundary. This happens whenever a Bragg condition is fulfilled, for special values of the total energy of the traveling particle, defining a band spectrum. This is the essence of Bloch theory for perfect metals. In such a case, the conductivity is infinite, if one neglects the influence of phonons and of the electron-electron interaction. If the medium is not periodic but quasiperiodic, such as quasicrystals, one may have also free Bloch waves if the quasiperiodic potential describing the influence of the ions on the travelling particle is not too strong [DiSi, BeLiTe, ChDe, BeIoScTe, BenSir].

However, in a disordered medium, the Bragg condition is unlikely, namely destructive interferences may force the electronic wave to vanish at infinity. Thus, the electonic wave is trapped in defects : in other words it is localized in a bounded region. Anderson proposed a tight binding model of such medium and could predict that 1-dimensional disordered chains always exhibit localization [Pas, Cyc]. Later on [AbAnLiRa] it was argued that in 2D the same effect occurs. But in higher dimension, localization holds only for strong disorder or at the band edges [FrSp, FrMaScSp]. Then if the disorder is not too strong, Ohm's law holds, leading to a finite conductivity, even if we ignore the phonons and the electron-electron interaction.

The Anderson model is extremely simple but contains most of the properties necessary to describe such a medium. In a tight binding representation, the electronic states can be represented as elements of the Hilbert space $\ell^2(\mathbf{Z}^D)$, if the crystal we start from is the D-dimensional lattice $\mathbf{Z}^D$. If there is no disorder, in the one electron approximation, the conduction electrons are approximately described by the free Laplacean Δ_D namely if $\psi \in \ell^2(\mathbf{Z}^D)$:

$$\Delta_D \psi(n) = t \sum_{|n-n'|=1} \psi(n') , \qquad (177)$$

where t is the "hopping" parameter which measures the energy required for an electron to hop from one site to the next one. The energy spectrum is then given by the band $[-2Dt, 2Dt]$.

Adding one defect in the crystal can be described by adding to the previous Laplacean a local potential in the form of a sequence $V_{defect} = (V_{defect}(n); n \in \mathbf{Z}^D)$, as was shown in 1949 by Slater. To get a homogeneous distribution of defects it is therefore sufficient to replace V_{defect} by a homogeneous sequence V. To take into account the randomness of the defect distribution we will assume that the values $V(n)$ of this potential at each site are identically distributed random variables. Even though we expect some correlation between them in realistic systems, at least at short distances, Anderson proposed to consider the simplest case for which they are independent and uniformly distributed in an interval $[-W, W]$. Then W is a measure of the disorder strength. Let Ω be the corresponding probability space (in this example, $\Omega = [-W, W]^{\mathbf{Z}^D}$) and let $\mathbf{P}$ be the corresponding probability measure (in this example, $\mathbf{P} = \otimes_{n \in \mathbf{Z}^D} dV(n)/2W)$. The potential becomes a function of the random variable $\omega \in \Omega$ so that the Anderson Hamiltonian can be written as :

$$H_\omega = \Delta_D + V_\omega . \tag{178}$$

The probability space $(\Omega, \mathbf{P})$ can be seen as the configuration space for the disorder. The translation invariance of the original lattice is not completely lost. For indeed, translating this new system is equivalent to translate the distribution of defects back. More precisely, there is a measure preserving action of the translation group on Ω. For the Anderson model this action is given by $T^r \omega_n = \omega_{n-r}$. If we denote by $T(r)$ the translation by $r \in \mathbf{Z}^D$ in the Hilbert space, namely for $\psi \in \ell^2(\mathbf{Z}^D)$, $T(r)\psi(n) = \psi(n-r)$, we get the following "covariance condition":

$$T(r)H_\omega T(r)^{-1} = H_{T^r \omega} . \tag{179}$$

We will complete this framework by adding two conditions. The first one is the ergodicity of the probability measure $\mathbf{P}$. Thanks to Birkhoff's ergodic theorem, it expresses the fact that space averages coincide with $\mathbf{P}$-average. In this way, $\mathbf{P}$ can be constructed in practice simply by taking space averages, an unambiguous process. The second one concerns the existence of a topology on Ω which makes it a compact Hausdorff space, and such that the $\mathbf{P}$-measurable sets are generated as a σ-algebra by the Borel sets, namely $\mathbf{P}$ is a Radon measure. In the Anderson model the product topology will do it. Actually an intrinsic definition of homogeneous system has been proposed in [Bel:Kth, Bel:Gap] leading to the definition of a canonical topology on the disorder configuration space. For this topology, the mapping $\omega \in \Omega \mapsto H_\omega$ is strongly continuous (in the resolvent sense whenever H_ω is unbounded self adjoint).

To summarize, homogeneous media, such as crystals, quasicrystals, glasses, amorphous, aperiodic or disordered systems, may be mathematically described by the following axioms.

(D1)-The disorder configuration space is a compact Hausdorff topological space Ω endowed with a probability Radon measure $\mathbf{P}$
(D2)-The translation group is a locally compact abelian group G acting in Ω by mean of a continuous group of homeomorphisms $\omega \mapsto g\omega$. The probability $\mathbf{P}$ is G-invariant and ergodic.
(D3)-The quantum state space is a separable Hilbert space $\mathcal{H}$ in which G acts through a projective unitary representation $\{T(g); g \in G\}$.
(D4)-The Hamiltonian is a strong-resolvent continuous family $H = \{H_\omega; \omega \in \Omega\}$ of self adjoint operators acting on $\mathcal{H}$ with a common G-invariant domain $\mathcal{D}$.
(D5)-A covariance condition is satisfied, namely:

$$T(g)H_\omega T(g)^{-1} = H_{g\omega} . \tag{180}$$

In general we will prefer a projective unitary representation. For indeed there are concrete examples for which the translation group does not act as a true representation. This is the case for a crystal in a uniform magnetic field [Bel:Gap]. We have restricted ourself to abelian translation groups because no concrete useful example have been studied till now with non abelian groups. However, systems living on a Cayley tree admits a non abelian translation group which is usually a free group. We can also include in G other symmetries like rotations, reflections, if necessary. This has never been investigated in detail yet, even though we believe that it should be useful: classification of defects in crystal may be related to such groups.

The smallest observable algebra that can be of interest for physics, is the one constructed with the energy. In more concrete systems however, other observables like spins, may be relevant. For simplicity, we will consider the simplest case for which the only relevant observable is the energy. In a homogeneous medium, the choice of the origin is arbitrary, since the systems reproduces itself under translation. So that the physics of the system is described by any of the elements of the family $H = \{H_\omega; \omega \in \Omega\}$ representing the energy. In order to avoid choosing arbitrarily one of them, we will include all of them. We then define a non commutative C^*-algebra $C^*(H)$ as the smallest one in the space of bounded operators on $\mathcal{H}$ containing the resolvent of each of the elements of H. In general, we do not know the structure of such an algebra. However for most concrete examples construct till now, namely by using the Schrödinger operator for one electron systems [Bel:Kth, Bel:Gap], like the Anderson model, this algebra is nothing but the crossed product $C(\Omega) \times G$ defined by the topological dynamical system (Ω, G) describing the disorder configurations in the original medium. This algebra must be slightly modified if a uniform magnetic field is turned on. We will ignore this latter case here.

Thanks to this framework, there is a very close analogy with aperiodic media in Solid State Physics and the dynamics of a kicked rotor. Even though the physical interpretation is very different, the C^*-algebra used to describe the observables is also a crossed product. However, in the kicked rotor model, the lattice G is the quantized momentum space instead, and the space Ω admits a fairly different interpretation since the variable γ plays the role of an effective Planck constant and is related to the period of the kicks, x plays the role of a magnetic field, whereas y represents a generic translation in momentum space. We also notice that the ergodicity of the measure holds only if $\gamma/2\pi$ is a fixed irrational number.

There is also a very close analogy with 2D-dimensional lattice electrons in a uniform magnetic field. We have already seen that the observable algebra is the rotation algebra $\mathcal{A}_I$ which can also be seen as the crossed product $C(I \times \mathbf{T}) \times_\phi \mathbf{Z}$ if $\phi : (\gamma, x) \in I \times \mathbf{T} \mapsto (\gamma, x + \gamma) \in I \times \mathbf{T}$. Then γ plays the role of a dimensionless magnetic flux per plaquette, whereas x is a generic position of the origin in the x-direction of the lattice. Again, the ergodicity of the measure on $\Omega = I \times \mathbf{T}$ holds only if $I = \{\gamma\}$ where $\gamma/2\pi$ is a fixed irrational.

The main question now is whether this formal analogy between so different problems will produce phenomena similar to Anderson's localization. The common belief is that if H is a selfadjoint operator belonging to this algebra, with short range interactions, namely if it is smooth enough with respect to the differential structure that will be described in the next subsection, it will exhibit such phenomena at least if the dynamical system (Ω, G) is "sufficiently aperiodic". The precise meaning of "sufficiently aperiodic" is not completely understood yet. Several numerical studies have investigate this question, but they are far from having given a precise criterion yet [FiHuXX]. More precisely we define a 2-point function by $C(g) = \langle F F_g \rangle - \langle F \rangle^2$, where F is a continuous function on Ω and $F_g(\omega) = F(g^{-1}\omega)$ while $\langle \cdot \rangle$ is the ergodic average. If any 2-point function converges to zero fast enough as $g \mapsto \infty$, the localization is expected to occur. This is certainly not the case for a periodic or an almost periodic dynamics, describing for instance a perfect crystal with or whithout a uniform magnetic field. And indeed we do not expect in this case localization to occur. Still, a 1D model like the Almost Mathieu Hamiltonian [AuAn, ChDe, BeLiTe], has been proved to exhibit a metal-insulator transition at large coupling. But the Furstenberg map for instance, which satisfies this criterion, should give rise to localization. This is the basis of an argument by Fishman, Grempel and Prange [FiGrPr] predicting that localization occurs in the kicked rotor problem.

The next problem therefore is to describe mathematically what we expect to characterize the localization. One of the first criterion used by Anderson was connected to the time evolution of quantum states : if the time-average of the probability for the initial state to come back after time t is positive, then localization do occur. We will see later on, thanks to an early result of Pastur [Pas] that this criterion is related to the existence of a point spectrum for H_ω, P-almost surely. This is essentially why mathematicians describe localization in term of the existence of a point spectrum. It is related to the finiteness of the so called "inverse participation ratio" (see below). Another way consists in defining the localization length: roughly speaking it gives a measure of the diameter of the region where a typical eigenstate is localized.

One of the main problems in dealing with the spectral property of the Hamiltonian, is that in many situations, this requires the choice of a fixed representation of the observable algebra. While in the Anderson model, this choice is quite natural, thanks to the description of the original disorders medium, in other models for which we would like to use the localization theory, it is not necessarily so. Two inequivalent representations of the same algebra may give different type of spectral measure for the same Hamiltonian. This happens for instance in the problem of Bloch electrons in a magnetic field. Therefore if this latter point of view were correct, localization would require to distinguish physically between different representations. However, the computation of the localization length requires a space average, in order to get a quantity insensitive to the specific configuration of the disorder, and therefore as we will see, it can be interpreted in a purely algebraic way. There is therefore an apparent contradiction between the two points of view. This is actually nothing but the usual opposition between the Schrödinger and Heisenberg point of view in Quantum Mechanics. Our main purpose in this section is to show how to reconcile them, and to show that in some sense they are equivalent.

Our last comment concerns the semiclassical limit. While this limit is meaningless in the Anderson problem, since the starting point is the band theory for perfect crystals, a fairly strong quantum theory, the kicked rotor problem gives a nice example where the semiclassical limit exists indeed together with a localization effect. It is therefore natural to consider what happens to the localization phenomena in this limit. The main discovery of Chirikov, Izrailev and Shepelyansky [ChIzSh] was to relate this limit to the diffusion constant in phase space of the classical kicked rotor. Even though this relation has not been proved to hold rigorously, many numerical studies show that it is probably correct at least under some unknown "generic condition". Therefore we have reached here one point of the so-called "quantum chaos". We will give only some pieces of this puzzle here.

7.2 The Observable Algebra

To avoid useless technical difficulties, we consider now the C^*-algebra $\mathcal{C}(\Omega) \times G$ where $G = \mathbf{Z}^D$. D will be called the dimension of the lattice. However, most of what will be described here can be extended to more general groups such as $\mathbf{R}^D$ for instance. As for the rotation or the Furstenberg algebra, we can develop a calculus as follows.

Elements of $\mathcal{C}(\Omega) \times \mathbf{Z}^D$ are continuous complex functions $a(\omega, n)$ on the space $\Omega \times \mathbf{Z}^D$ vanishing at infinity. To define this algebra properly, it is more convenient to start with the dense subalgebra $\mathcal{C}_c(\Omega \times \mathbf{Z}^D)$ of continuous functions on $\Omega \times \mathbf{Z}^D$ with compact support, endowed with the following operations:

$$ab(\omega; n) = \sum_{l \in \mathbf{Z}^D} a(\omega; l) b(T^{-l}\omega; n - l) , \tag{181}$$

$$a^*(\omega; n) = \overline{a(T^{-n}\omega; -n)} . \tag{182}$$

Since the functions a and b have compact support, the sum above is finite. Remarkable elements are given by :

$$\mathbf{I}(\omega; n) = \delta_{n,0} , \qquad U(r)(\omega; n) = \delta_{n,-r} , \qquad r \in \mathbf{Z}^D . \tag{183}$$

The first one $\mathbf{I}$ is a unit, whereas $U(r)$ is a group of unitaries namely $U(r)U(r') = U(r + r')$, $U(0) = \mathbf{I}$ and $U(r)^* = U(-r) = U(r)^{-1}$.

A family of representations in the Hilberts space $\ell^2(\mathbf{Z}^D)$ indexed by $\omega \in \Omega$ is given by:

$$\pi_\omega(a)\psi(n) = \sum_{n' \in \mathbf{Z}^D} a(T^{-n}\omega; n' - n)\psi(n') , \qquad a \in \mathcal{C}_c(\Omega \times \mathbf{Z}^D) , \qquad \psi \in \ell^2(\mathbf{Z}^D) . \tag{184}$$

In particular we get $a(\omega; n) = \langle 0|\pi_\omega(a)|n\rangle$. Then a C^*-norm is defined by:

$$\|a\| = \sup_{\omega \in \Omega} \|\pi_\omega(a)\| , \qquad a \in \mathcal{C}_c(\Omega \times \mathbf{Z}^D) . \tag{185}$$

Then $\mathcal{C}(\Omega) \times_T \mathbf{Z}^D$ is the completion of $\mathcal{C}_c(\Omega \times \mathbf{Z}^D)$ under this norm. To shorten the notations we will denote it by $\mathcal{A}$.

Given an invariant probability measure $\mathbf{P}$ on Ω, a normalized trace $\tau_{\mathbf{P}}$ (or τ for short whenever no confusion arises) is defined by:

$$\tau(a) = \int_\Omega d\mathbf{P} a(\omega; 0) , \qquad a \in \mathcal{A} . \tag{186}$$

It is easy to see, by using the Birkhoff ergodic theorem, that if $\mathbf{P}$ is ergodic,

$$\tau(a) = \lim_{\Lambda \uparrow \mathbf{Z}^D} \frac{1}{|\Lambda|} \mathrm{Tr}_\Lambda (\pi_\omega(a)) , \qquad \text{for } \mathbf{P} - \text{almost all } \omega . \tag{187}$$

At last, the differential structure is related to the group action and defined as follows. If $n = (n_1, \ldots, n_D) \in \mathbf{Z}^D$, we define the $*$-derivation ∂_μ by:

$$\partial_\mu a(\omega; n) = in_\mu a(\omega; n) , \qquad \mu = 1, \ldots, D . \tag{188}$$

These derivations commute together and are the infinitesimal generators of the D-parameter group of automorphism $\{\rho_\theta; \theta_1 \mathbf{T}^D\}$ (the so-called dual action of Takasaki [Ped]) defined by:

$$\rho_\theta(a)(\omega; n) = e^{i\theta n} a(\omega; n) , \qquad \theta n = \theta_1 n_1 + \ldots + \theta_D n_D . \qquad (189)$$

Moreover, denoting by N_μ the position operators defined by $N_\mu \psi(n) = n_\mu \psi(n)$ in $\ell^2(\mathbf{Z}^D)$, we get:

$$\pi_\omega(\partial_\mu a) = i[N_\mu, \pi_\omega(a)] . \qquad (190)$$

7.3 Localization Criteria

In this subsection we give several criteria for the localization and discuss the relation between its finiteness and the nature of the spectrum. We will consider a self adjoint element $H = H^*$ in the algebra $\mathcal{A} = \mathcal{C}(\Omega) \times_T \mathbf{Z}^D$ previously described. In view of the study of a Floquet operator we may consider a unitary element $F = (F^*)^{-1}$ of this algebra instead. This latter case reduces to the former provided we identify F with e^{iTH} for some $T > 0$ and the Borel sets Δ are subset of the unit circle. In the physical representation π_ω we consider the operator $\pi_\omega(H) = H_\omega$ instead.

If Δ is some Borel subset of $\mathbf{R}$ we denote by P_Δ t eigenprojection of H corresponding to energies in Δ namely :

$$P_\Delta = \chi_\Delta(H) , \qquad (191)$$

where χ_Δ is the characteristic function of the interval Δ. Again, we notice that in general P_Δ may not belong to $\mathcal{A}$. However it always belongs to the so-called Borel algebra $\mathcal{B}(\mathcal{A})$ [Ped], formally generated by Borel functions of elements of $\mathcal{A}$. The Borel functional calculus permits to extend any representation of $\mathcal{A}$ to its Borel algebra. Hence the previous definition makes sense. The price we pay for it is that the mapping $\omega \in \Omega \mapsto \pi_\omega(a)$ may not necessarily be strongly continuous any more, but it is always strongly borelian if $a \in \mathcal{B}(\mathcal{A})$.

If H_ω has a pure point spectrum in Δ we get the following decomposition:

$$\pi_\omega(P_\Delta) = \sum_{E \in \Delta} \Pi_E(\omega) , \qquad (192)$$

where $\Pi_E(\omega)$ is the eigenprojection of H_ω corresponding to the eigenvalue E. If E is a simple eigenvalue, one gets $\Pi_E(\omega) = |\psi_{E,\omega}\rangle\langle\psi_{E,\omega}|$, where $\psi_{E,\omega}$ is a normalized eigenstate namely:

$$\|\psi_{E,\omega}\|^2 = \sum_{n \in \mathbf{Z}^D} |\psi_{E,\omega}(n)|^2 = 1 < +\infty . \qquad (193)$$

The first quantity measuring the localization is the probability of staying at the origin. It was introduced by [And] and studied by Pastur [Pas]. To define it let us first consider the time-average $A_{n,n'}(\Delta, \omega)$ of the probability for an initial state at n to be localized at n' after time t:

$$A_{n,n'}(\Delta, \omega) = \lim_{T \to \infty} \int_0^T \frac{dt}{T} |\langle n|\pi_\omega(e^{itH} P_\Delta)|n'\rangle|^2 . \qquad (194)$$

If H_ω has a pure point spectrum, the decomposition (192) leads to :

$$A_{n,n'}(\Delta, \omega) = \sum_{E \in \Delta} |\langle n|\Pi_E(\omega)|n'\rangle|^2 . \tag{195}$$

The covariance condition implies $A_{n,n'}(\Delta, \omega) = A_{0,n'-n}(\Delta, T^{-n}\omega)$, so that the staying probability is entirely given by the function $A_{0,0}(\Delta, \omega)$, provided we consider it as a function of the disorder. We remark that if the eigenvalues are simple, since the eigenstates are normalized we get:

$$A_{0,0}(\Delta, \omega) = \frac{\sum_{E \in \Delta} |\psi_{E,\omega}(0)|^4}{\left(\sum_{E \in \Delta} |\psi_{E,\omega}(0)|^2\right)^2} , \tag{196}$$

namely $A_{0,0}(\Delta, \omega)$ is the mean inverse participation ratio for energies in Δ. To get a quantity insensitive to the disorder, let us average it with respect to $\mathbf{P}$ defining the averaged inverse participation ratio :

$$\xi_\Delta = \int_\Omega d\mathbf{P} \, A_{0,0}(\Delta, \omega) . \tag{197}$$

Using now the automorphism group defined in eq.(189) and the eq.(184,186), an elementary calculation leads to the following expression for ξ_Δ:

$$\xi_\Delta = \lim_{T \to \infty} \int_0^T \frac{dt}{T} \int_{\mathbf{T}^D} \frac{d^D\theta}{(2\pi)^D} \, \tau(e^{itH} P_\Delta \rho_\theta(e^{-itH} P_\Delta)) . \tag{198}$$

So we see that the staying probability or the inverse participation ratio, admits a purely algebraic expression. The Pastur theorem [Pas] can then be established as follows:

Theorem 13 *For almost all $\omega \in \Omega$, the number of eigenvalues of H_ω in Δ is either zero or infinity. The latter is realized, namely H_ω has some point spectrum in Δ, if and only if the averaged inverse participation ratio ξ_Δ is positive.*

Comment: this criterion is not sufficient to eliminate continuous spectrum.

We now introduce a stronger notion of localization giving a measurement of the localization length. Whenever $\pi_\omega(H)$ has pure point spectrum, the eigenstate may decay faster at infinity on the lattice. We are led to introduce quantities like:

$$\ell^{(p)}(E, \omega) = \left[\sum_{n \in \mathbf{Z}^D} |\psi_{E,\omega}(n)|^2 |n|^p \right]^{1/p} , \tag{199}$$

for $p \geq 1$. If the eigenstates decrease exponentially fast one can also consider the quantity

$$\ell(E, \omega) = \limsup_{n \to \infty} \frac{-ln|\psi_{E,\omega}(n)|}{|n|} . \tag{200}$$

However such expressions are very badly behaving with ω in general and they are not suited for comparison with experiments or numerical calculations. The following definition will be more convenient and will give rise to a quantity independent of ω. We consider the averaged fluctuation of the position in the form:

$$\Delta X_{\omega,n}(T)^2 = \int_0^T \frac{dt}{T} \langle n|(N_\omega(t) - N)^2|n\rangle , \qquad (201)$$

where $N_\omega(t) = e^{iH_\omega t} N e^{-iH_\omega t}$, and $N = (N_1, \ldots, N_D)$ is the position operator. The covariance property gives $\Delta X_{\omega,n}(T) = \Delta X_{T-n\omega,0}(T)$, so that after averaging over the disorder, we get a quantity independent of n, namely $\Delta X(T)^2 = \int_\Omega d\mathbf{P}(\omega) \Delta X_{\omega,0}(T)^2$. An elementary calculation shows that:

$$\Delta X(T)^2 = \int_0^T \frac{dt}{T} \tau(|\nabla e^{iHt}|^2) . \qquad (202)$$

We will generalize this expression by considering, for every Borel subset Δ of the real line, the corresponding quantity $\Delta X_\Delta(T)^2$ obtained in the same way if we replace e^{iHt} by $e^{iHt} P_\Delta$. The main result in this respect is the following:

Theorem 14 *If*

$$\ell^2(\Delta) = \limsup_{T \to \infty} \Delta X_\Delta(T)^2 < \infty , \qquad (203)$$

then H_ω has a pure point spectrum in Δ for almost all $\omega \in \Omega$. Moreover, if $\mathcal{N}$ denotes the density of states of H, there is an $\mathcal{N}$-measurable non negative function ℓ on $\mathbf{R}$ such that for every Borel subset Δ' of Δ,

$$\ell^2(\Delta') = \int_{\Delta'} d\mathcal{N}(E)\ell(E)^2 , \qquad (204)$$

Comment: we will see in the proof that if $\sigma_{pp}(\omega)$ denotes the set of eigenvalues of H_ω :

$$\ell^2(\Delta') = \int_\Omega d\mathbf{P}(\omega) \sum_{n \in \mathbf{Z}^D} n^2 \sum_{E \in \sigma_{pp}(\omega) \cap \Delta} |\langle 0|\Pi_E(\omega)|n\rangle|^2 . \qquad (205)$$

In particular, letting Δ shrink to the point E, the function $\ell(E)^2$ represents a kind of average (over the disorder and over a small spectral set around E), of the quantity $\sum_{n \in \mathbf{Z}^D} n^2 |\psi_{E,\omega}(n)|^2$. Namely it is a measure of the extension of the eigenstate corresponding to E. This is the reason for the definition below.

Definition 2 *The function ℓ will be called the localization length for H.*

Proof of the theorem: (i)-The basic argument we will use here is due to Guarneri [Gua, Bel:Tre]. We will denote by $\sigma_{pp}(\omega)$ the set of eigenvalues of H_ω (the point spectrum), whereas $\Pi_{pp}(\omega)$ will denote its spectral projection on the point spectrum and $\Pi_c(\omega) = \mathbf{I} - \Pi_{pp}(\omega)$ will be its spectral projection on the continuous part of its spectrum. Using the definition of the trace in $\mathcal{A}$, we get :

$$\Delta X_\Delta(T)^2 = \int_\Omega d\mathbf{P}(\omega) \sum_{n \in \mathbf{Z}^D} n^2 p_T(\omega, n) , \qquad (206)$$

where,

$$p_T(\omega, n) = \int_0^T \frac{dt}{T} |\langle 0|\pi_\omega(e^{iHt} P_\Delta)|n\rangle|^2 . \qquad (207)$$

We will set $p_T(n) = \int_\Omega d\mathbf{P}(\omega) p_T(\omega, n)$. By definition, we have :

$$0 \leq p_T(\omega, n) \leq 1 , \tag{208}$$

$$\sum_{n \in \mathbf{Z}^D} p_T(\omega, n) = 1 , \tag{209}$$

whereas the Wiener criterion gives

$$\lim_{T \mapsto \infty} p_T(\omega, n) = \sum_{E \in \sigma_{pp}(\omega) \cap \Delta} |\langle 0|\Pi_E(\omega)|n\rangle|^2 . \tag{210}$$

In particular, if L is a positive integer, and $|n|_\infty = \max_{1 \leq j \leq D} |n_j|$,

$$\lim_{T \mapsto \infty} \sum_{|n|_\infty < L} p_T(\omega, n) \leq \langle 0|\Pi_{pp}(\omega)|0\rangle = 1 - \langle 0|\Pi_c(\omega)|0\rangle . \tag{211}$$

If we set $r = \int_\Omega d\mathbf{P}(\omega) \langle 0|\Pi_c(\omega)|0\rangle$ we obtain after averaging over the disorder

$$\lim_{T \mapsto \infty} \sum_{|n|_\infty < L} p_T(n) \leq 1 - r . \tag{212}$$

Since $r \geq 0$, one can find $T_L > 0$ such that if $T \geq T_L$, $\sum_{|n|_\infty < L} p_T(n) \leq 1 - r/2$. Thus :

$$\Delta X_\Delta(T)^2 \geq L^2 \int_\Omega d\mathbf{P}(\omega) \sum_{|n|_\infty \geq L} p_T(\omega, n) \geq L^2 (1 - \sum_{|n|_\infty < L} p_T(n)) \geq \frac{L^2 r}{2} . \tag{213}$$

Taking the limit $T \mapsto \infty$ leads to $L^2 r \leq 2\ell^2(\Delta) < \infty$ for any $L \in \mathbf{N}$. Thus $r = 0$ showing that for almost all ω's, $\langle 0|\Pi_c(\omega)|0\rangle = 0$. Using the covariance condition we also get for all n's $\langle n|\Pi_c(\omega)|n\rangle = 0$ almost surely, and since $\mathbf{Z}^D$ is countable, there is $\Omega' \subset \Omega$ of probability one such that for any $\omega \in \Omega'$, $\Pi_c(\omega)|n\rangle = 0$ for all $n \in \mathbf{Z}^D$, namely the continuous spectrum is empty.

(ii)-Given two Borel subsets $\Delta_1, \Delta_2 \subset \Delta$, we define the following expression:

$$\mathcal{E}_{T,\omega}^{(L)}(\Delta_1, \Delta_2) = \int_0^T \frac{dt}{T} \sum_{|n|_\infty < L} n^2 \langle 0|\pi_\omega(e^{iHt} P_{\Delta_1})|n\rangle \overline{\langle 0|\pi_\omega(e^{iHt} P_{\Delta_2})|n\rangle} . \tag{214}$$

In particular $\mathcal{E}_{T,\omega}^{(L)}(\Delta, \Delta) = \sum_{|n|_\infty < L} n^2 p_T(\omega, n)$. This expression gives a Borel function of ω. In addition, using the Wiener criterion, we have:

$$\lim_{T \mapsto \infty} \mathcal{E}_{T,\omega}^{(L)}(\Delta_1, \Delta_2) = \sum_{|n|_\infty < L} n^2 \sum_{E \in \sigma_{pp}(\omega) \cap \Delta_1 \cap \Delta_2} |\langle 0|\Pi_E(\omega)|n\rangle|^2 = \mathcal{E}_\omega^{(L)}(\Delta_1 \cap \Delta_2) . \tag{215}$$

From this definition of $\mathcal{E}_\omega^{(L)}(\Delta')$ whenever $\Delta' \subset \Delta$ is Borel, it follows that

(a) $0 \leq \mathcal{E}_\omega^{(L)}(\Delta') \leq L^2$,
(b) If $\Delta_1 \cap \Delta_2 = \emptyset$, then $\mathcal{E}_\omega^{(L)}(\Delta_1 \cup \Delta_2) = \mathcal{E}_\omega^{(L)}(\Delta_1) + \mathcal{E}_\omega^{(L)}(\Delta_2)$,
(c) If $(\Delta_i)_{i \in \mathbf{N}}$ is a decreasing sequence of Borel susbets of Δ converging to the empty set, namely $\bigcap_{i \in \mathbf{N}} \Delta_i = \emptyset$, then $\mathcal{E}_\omega^{(L)}(\Delta_i)$ decreases to zero,
(d) $\mathcal{E}_\omega^{(L)}(\Delta') \leq \mathcal{E}_\omega^{(L+1)}(\Delta')$,
(e) $\mathcal{E}_\omega^{(L)}(\Delta')$ is a Borel function of ω as a pointwise limit of Borel functions.

After averaging over the disorder we obtain $\mathcal{E}^{(L)}(\Delta') = \int_\Omega d\mathbf{P}(\omega)\mathcal{E}_\omega^{(L)}(\Delta')$ which fulfill (a), (b), (c) using the dominated convergence theorem, and (d). From (202,206), we also get:

$$\int_\Omega d\mathbf{P}(\omega)\mathcal{E}_{T,\omega}^{(L)}(\Delta',\Delta') \le \int_0^T \frac{dt}{T}\tau(|\nabla e^{iHt}P_{\Delta'}|^2) . \qquad (216)$$

Using the dominated convergence theorem we conclude that $\mathcal{E}^{(L)}(\Delta') \le \ell^2(\Delta')$ and also thanks to the property (b), $\mathcal{E}^{(L)}(\Delta') \le \ell^2(\Delta)$ for $\Delta' \subset \Delta$. It follows that $\lim_{L\to\infty} \mathcal{E}^{(L)}(\Delta') = \mathcal{E}(\Delta')$ exists and defines a non negative σ-additive set function over the set of Borel subsets of Δ, namely a Radon measure. Moreover it satisfies $\mathcal{E}(\Delta') \le \ell^2(\Delta')$, and by the monotone convergence theorem, eq.(215) above implies:

$$\mathcal{E}(\Delta') = \int_\Omega d\mathbf{P}(\omega) \sum_{n\in\mathbf{Z}^D} n^2 \sum_{E\in\sigma_{pp}(\omega)\cap\Delta} |\langle 0|\Pi_E(\omega)|n\rangle|^2 . \qquad (217)$$

(iii)-On the other hand the definition of $\ell^2(\Delta')$ and Fatou's lemma imply:

$$\ell^2(\Delta') \le \int_\Omega d\mathbf{P}(\omega) \sum_{n\in\mathbf{Z}^D} n^2 \limsup_{T\to\infty} \int_0^T \frac{dt}{T}|\langle 0|\pi_\omega(e^{iHt}P_{\Delta'})|n\rangle|^2 . \qquad (218)$$

By the Wiener criterion the right hand side is nothing but $\mathcal{E}(\Delta')$ showing that $\ell^2(\Delta') = \mathcal{E}(\Delta') \le \ell^2(\Delta)$ for $\Delta' \subset \Delta$. Hence it is a nonnegative Radon measure on Δ.
(iv)-To finish the proof it is sufficient to show that this measure is absolutely continuous with respect to the DOS $\mathcal{N}$ of H. Let then $\Delta' \subset \Delta$ be such that $\mathcal{N}(\Delta') = \tau(\Delta') = 0$. From the definition of the trace it follows that $\langle 0|\pi_\omega(P_{\Delta'})|0\rangle = 0$ almost surely. By covariance and because $\mathbf{Z}^D$ is countable this gives $\pi_\omega(P_{\Delta'})|n\rangle = 0$ for all n's almost surely, namely $\pi_\omega(P_{\Delta'}) = 0$ almost surely. Then (214) above implies $\mathcal{E}_{T,\omega}^{(L)}(\Delta',\Delta') = 0$ for any L, T, and almost every ω. Consequently $\mathcal{E}(\Delta') = 0$, and the representation (204) holds.

7.4 Localization in the Kicked Rotor

As claimed previously, one can use the same formalism for investigating the localization properties of the kicked rotor. It is then sufficient to work with the Floquet operator instead of a Hamiltonian, and with Borel subsets of the circle. However the C^*-algebra we are using, $\mathcal{B}_\gamma$, is parametrized by the effective Planck constant γ, an additional parameter here. Apart from this remark, we get the previous structure if we set $\Omega = \mathbf{T}^2$, $D = 1$, and the action is provided by the Furstenberg map. The Lebesgue measure $dxdy/4\pi^2$ gives the probability measure, which is ergodic whenever $\gamma/2\pi$ is irrational. In view of the theorem (14) above, we cannot expect any finite localization length otherwise, because the action is no longer ergodic and from a result of Izrailev and Shepelyanski [IzSh] it follows that we get an absolutely continuous spectrum for $\pi_{\gamma,x,y}(F)$ whenever $\gamma/2\pi$ and $x/2\pi$ are rational. Now, we remark that the definition of the localization length coincides with the definition of the mean kinetic energy given by (176) up to the constant γ^2. Hence $F_{K,\gamma,x}$ will have a pure point spectrum in Δ whenever the mean kinetic energy

$$\overline{\mathcal{E}}_\Delta(\gamma) = \gamma^2 \limsup_{T\to\infty} \frac{1}{T}\sum_{t=0}^{T-1} \tau_\gamma\left(|\partial_\theta(F_{K,\gamma}^t P_\Delta)|^2\right) , \qquad (219)$$

is finite. Moreover we get the elementary formula $\gamma^2 \ell_\gamma^2(\Delta) = \overline{\mathcal{E}}_\Delta(\gamma)$ whenever ℓ_γ denotes the localization length in term of γ.

The case of the kicked rotor permits to go a little bit further. First of all, the definition of the Floquet operator permits to show that it is C^∞ with respect to ∂_θ. Moreover we get the folowing result:

Proposition 10 *If the localization length ℓ_γ exists for the Floquet operator $F_{K,\gamma}$ of the kicked rotor model, it is constant over the circle.*

Proof: Clearly $\eta_y = \hat{\rho}_{\theta=0,y}$ (see (135)) commutes with the derivation ∂_θ. Moreover η_y translates the spectrum of $F_{K,\gamma}$ by y along the circle because for any Borel subset Δ of $\mathbf{T}$:

$$\eta_y(F_{K,\gamma}) = e^{iy} F_{K,\gamma} , \qquad \eta_y(P_\Delta) = P_{\Delta+y} . \tag{220}$$

Thus

$$\tau_\gamma \left(|\partial_\theta (F_{K,\gamma}^t P_\Delta)|^2 \right) = \tau_\gamma \left(\eta_y |\partial_\theta (F_{K,\gamma}^t P_\Delta)|^2 \right) = \tau_\gamma \left(|\partial_\theta (F_{K,\gamma}^t P_{\Delta+y})|^2 \right) , \tag{221}$$

because η_y is an automorphism. It implies $\ell_\gamma^2(\Delta) = \ell_\gamma^2(\Delta + y)$ for any Borel set Δ, and therefore $\ell(E) = $ const. Thus Δ is not needed anymore so that :

Corollary 5 *For the kicked rotor model the following formula holds*

$$\ell_\gamma^2 = \frac{\overline{\mathcal{E}}_\gamma}{\gamma^2} = \lim_{T \mapsto \infty} \frac{1}{T} \sum_{t=0}^{T-1} \tau_\gamma \left(|\partial_\theta (F_K^t)|^2 \right) \tag{222}$$

A result by Casati & Guarneri [CaGu] shows that, the spectral measure of $F_{K,\gamma,x}$ is purely continuous generically in γ. Thus :

Proposition 11 *For the kicked rotor model there is a dense G_δ-set Γ of zero Lebesgue measure in $[0,1]$ such that for any $\gamma \in \Gamma$ the localization length diverges.*

However many numerical calculations [CaChIzFo, BeBa] have shown that the mean kinetic energy for the quantum kicked rotor model is bounded in time. So we expect the localization length to be finite on a "large set" of γ's, presumably for almost all γ's in $[0,1]$. Before discussing this question let us mention without proof another result which supplement the previous one namely

Proposition 12 *For the kicked rotor model the localization length is a lower semi-continuous function of γ.*

We may also expect $\gamma^2 \ell^2(\gamma)$ to converges to some finite quantity as $\gamma \mapsto 0$. This is the content of the Chirikov-Izrailev-Shepelyansky formula [ChIzSh] found on the basis of a numerical work. The well-known observation is that despite the diffusive behavior of the classical model (namely for strong coupling) the quantized version exhibits, up to a certain breaking time τ^*, a diffusion-like motion in phase space and then for $t > \tau^*$ its kinetic energy saturates as a function of time. This numerical result allows us to write

$$\overline{\mathcal{E}}_\gamma = \mathcal{E}_\gamma(\tau^*) \simeq D\tau^* , \tag{223}$$

where D is the classical diffusion coefficient.

There is here a mathematical difficulty. First of all, never was a diffusion coefficient shown to exist rigorously for the standard map. Moreover, averaging it over all possible initial conditions will not give a finite quantity due to the "Pustilnikov acceleration modes" (or "islands of stability"). This means that we should not average over the full torus. It raises the question of which quantum average should be considered. However, recent works [BeVa, Vai, Cher] have shown that for the sawtooth map, a diffusion coefficient does exist. Moreover, a conjecture states that for the standard map there is a "large" set of values of K for which no island of stability occurs, and a diffusion constant does exist.

To get the Chirikov-Izrailev-Shepelyansky formula, we argue as follows. Since the eigenstates of the Floquet operator are localized, only a finite number $\ell \simeq \ell_\gamma$ of eigenvalues contribute effectively to the evolution of the initial state $|0\rangle$. Therefore we can approximate this Floquet operator by a $\ell \times \ell$ matrix $F^{(\ell)}$. The existence of classical chaos will lead to a strong level repulsion. Hence one can consider that the mean distance between the quasienergies is $\Delta E \approx 2\pi/\ell = O(\gamma)$ on the torus.

For times short enough, the discrete spectral sum arising from the previous approximation can be approximated by an integral, which will be precisely the classical approximation. Hence for t small, $\mathcal{E}_\gamma(t) \approx \mathcal{E}_{\text{cl}}(t) \approx Dt$. This is fine as long as $t\Delta E \ll 2\pi$. But after a breaking time $\tau^* \approx 2\pi/\Delta E \approx \ell$, the quantization dominates and gives an almost periodic function of time for $\mathcal{E}_\gamma(t)$. Thus, $\ell_\gamma^2 \gamma^2 = \overline{\mathcal{E}_\gamma} \simeq \mathcal{E}_\gamma(\tau^*) \simeq D\tau^*$. Since $\ell \simeq \ell_\gamma$, we get :

$$\ell_\gamma \simeq \frac{D}{\gamma^2} \ . \tag{224}$$

Numerical calculations are in a fairly good agreement with this prediction, but no rigorous mathematical work has been produced to justify this formula yet. We may expect that :

$$\lim_{\gamma \to 0} \ell_\gamma \gamma^2 = D \quad \text{at } K \text{ large} , \tag{225}$$

under certain conditions. For indeed, we have seen that ℓ_γ diverges on a generic set of γ's. Moreover, D does not exist for all K's.

For the moment we do not know how to define mathematically the breaking time τ^*.

We would like now to study the behavior of the kinetic energy for the quantized version of the kicked rotor model as the effective Planck constant γ tends to zero. For that, we perform a numerical calculation giving the classical and quantum energies of the KR for $K = 4$ corresponding to the diffusive regime (Fig. 11). We computed the quantal energy for different values of Planck's constant γ in both cases; it is easy to see that as γ is decreased the quantal curves tend to the classical one.

One could think that this energy converges to its classical limit as $\gamma \mapsto 0$ but a problem arises because of the uniformity of the semiclassical limit with respect to time.

That the breaking time be $O(\gamma^2)$ can be shown by the following heuristic argument [HeTo]. The semiclassical approximation [Gut:Hou] for the evolution is correct modulo error terms of $O(\hbar(\gamma^2))$. Therefore, the quantum and classical evolutions for observables should agree up to time $O(\hbar^{-2})$. Whenever the semiclassical approximation is exact, however, such as in the hydrogen atom, the harmonic oscillator, the Arnold Cat map, we should not see any breaking time.

Appendix

Our aim in this appendix is to prove the theorem 1. The following theorem 15 actually implies the theorem 1.

Let $\mathcal{H}$ be a separable Hilbert space and H be a self adjoint operator on $\mathcal{H}$ with domain $\mathcal{D}$. This domain becomes a Hilbert space when endowed with the norm $\|\psi\|_H^2 = \|\psi\|^2 + \|H\psi\|^2$. Let also V be a bounded self adjoint operator on $\mathcal{H}$ leaving the domain $\mathcal{D}$ invariant and bounded on it for the domain norm $\|\cdot\|_H$. Let also f be a periodic continuous function on $\mathbf{R}$ with period T. Then the solution of the Schrödinger equation :

$$i\hbar\psi_t = (H + f(t)V)\psi_t , \tag{226}$$

with $\psi(s) = \psi$, is given by

$$\psi_t = U(t,s)\psi , \tag{227}$$

where $U(t,s)$ is a unitary operator such that :

(i) it is strongly continuous with respect to s, t ,
(ii) $U(s,s) = \mathbf{I}$ for all $s \in \mathbf{R}$,
(iii) $U(t,s) = U(t,t')Ut',s)$ for all $t' \in \mathbf{R}$,
(iv) $U(t+T,s+T) = U(t,s)$ for all $s,t \in \mathbf{R}$,
(v) for any $\psi \in \mathcal{D}$, the vector $U(t,s)\psi$ belongs to $\mathcal{D}$, is strongly differentiable with respect to s and t and is a solution of the Schrödinger equation.

The operator $F_s = U(s+T,s)$ is called the Floquet operator for the family $H(t) = H + f(t)V$. Notice that if $t = s$ the corresponding Floquet operators are unitarily equivalent thanks to (iii) and (iv).

Now for ϵ a positive real number, let ρ_ϵ be a non negative function on $\mathbf{R}$ with support in the interval $[-\epsilon, \epsilon]$ and of integral equal to one. We will set :

$$f_\epsilon(t) = \sum_{n\in\mathbf{Z}} \rho_\epsilon(t - nT). \tag{228}$$

Let F_ϵ be the corresponding Floquet operator with $t = -\epsilon$. Then the following result holds :

Theorem 15 *As ϵ tends to zero, the Floquet operator F_ϵ converges strongly to the unitary operator F given by :*

$$F = e^{-i(\frac{TH}{\hbar})}.e^{-i(\frac{V}{\hbar})} . \tag{229}$$

Proof : Denoting by $U_\epsilon(t,s)$ the evolution operator, it is a classical result that it admits the following Dyson expansion, which converges in norm :

$$U_\epsilon(t,s) = \sum_{n\geq 0} \left(-\frac{i}{\hbar}\right)^n \int_{s\leq s_n \leq \ldots \leq s_1 \leq t} ds_1 \ldots ds_n f_\epsilon(s_1)f_\epsilon(s_2)\ldots f_\epsilon(s_n)$$
$$e^{-(t-s_1)\frac{iH}{\hbar}}Ve^{-(s_1-s_2)\frac{iH}{\hbar}}V\ldots Ve^{-(s_n-s)\frac{iH}{\hbar}} . \tag{230}$$

Each term is a well defined strong integral. Taking $t = T - \epsilon$ and $s = -\epsilon$ we get an expansion for the Floquet operator.

As ϵ tends to zero, the restriction of the measure $f_\epsilon(s)ds$ to the interval $[-\epsilon, T-\epsilon]$ converges weakly to the Dirac measure supported by $\{0\}$. Since the integrand is strongly continuous, the term of order n in the Dyson expansion of F_ϵ converges strongly to $\left(-\frac{i}{\hbar}\right)^n e^{-i\left(\frac{HT}{\hbar}\right)} V^n/n!$. Summing up all these terms gives the result.

References

[AbAnLiRa]   E. Abrahams, P.W. Anderson, D.C. Licciardello, T.V. Ramakrishnan, Phys. Rev. Lett. **42** (1979) 673.

[And]   P.W. Anderson, *Absence of Diffusion in Certain Random Lattices*, Phys. Rev. **109** (1958) 1492-1505; *Local Moments and Localized States*, Rev. of Mod. Phys. **50** (1978) 191-201.

[AuAn]   S. Aubry, G. André, *Ann. Israeli. Phys. Soc.* **3**, 133 (1980).

[AvSi]   J. Avron, P.H.M. van Mouche, B. Simon, *On the measure of the spectrum for the Almost Mathieu Operator*, to be published in Commun. Math. Phys (1992).

[BaBeFl]   A. Barelli, J. Bellissard, R. Fleckinger, in preparation (1993)

[BaFl]   A. Barelli, R. Fleckinger, *Semiclassical analysis of Harper-like models*, to be published in Phys. Rev. B (1992)

[BaKr]   A. Barelli, C. Kreft, *Braid structure in a Harper model as an example of phase space tunneling*, J. Phys. I France **1** (1991) 1229-1249.

[Bel:Tre]   J. Bellissard, *Stability and Instability in Quantum Mechanics*, dans *Trends and Developments in the 80's*, S. Albeverio & P. Blanchard Eds., World Scientific, Singapour (1985).

[Bel:Kth]   J. Bellissard, *K-theory of C*-algebras in Solid State Physics*, Lect. Notes in Phys. **257**, Springer, Berlin, Heidelbe New York (1986).

[Bel:Eva]   J. Bellissard, *C*-Algebras in Solid State Physics : 2D electrons in a uniform magnetic field* in *Operator Algebras and Application*, Vol. 2, D.E. Evans & M. Takesaki Eds., Cambridge University Press (1988).

[Bel:Gap]   J. Bellissard, *Gap labelling theorems for Schrödinger's operators*, in *Number Theory and Physics*, J.-M. Luck, P. Moussa and M. Waldschmidt Eds., Springer Proceedings in Physics **47**, Springer, Berlin, Heidelberg, New York (1993).

[BeBa]   J. Bellissard, A. Barelli, *Dynamical Localization : Mathematical Framework*, in *Quantum Chaos, Quantum Measurement*, P. Cvitanovic, I.C. Percival, A. Wirzba Eds., Kluwer Publ. (1992) 105-129.

[BeBoGh]   J. Bellissard, A. Bovier, J.-M. Ghez, *Gap labelling theorems for one dimensional discrete Schrödinger operators*, Rev. Math. Phys. 4 (1992) 1-37.

[BeIoScTe]   J. Bellissard, B. Iochum, E. Scoppola, D. Testard, *Spectral properties of one dimensional quasi-crystals*, Commun. Math. Phys. **125** (1989) 527-543.

[BeKrSe]   J. Bellissard, C. Kreft, R. Seiler, *Analysis of the spectrum of a particle on a triangular lattice with two magnetic fluxes by algebraic and numerical methods*, J. Phys. A **24** (1991) 2329-2353.

[BeLiTe]   J. Bellissard, R. Lima, D. Testard, *Almost periodic Schrödinger Operators* in Mathematics + Physics, Lectures on Recent Results, Vol. **1**, L. Streit Ed., World Scientific, Singapore, Philadelphia (1985) 1-64.

[BeVa]   J. Bellissard, S. Vaienti, *Rigorous Diffusion Properties for the Sawtooth Map*, Commun. Math. Phys. **144** (1992) 521-536.

[BeVi]   J. Bellissard, M. Vittot, *Heisenberg's picture and non commutative geometry of the semiclassical limit in quantum mechanics*, Ann. Inst. H. Poincaré *52* (1990) 175-235.

[BenSir]   V.G. Benza, C. Sire, Phys. Rev. B **44** (1991) 10343.

[Bou]   N. Bourbaki, *Théories Spectrales*, Hermann, Paris (1967).

[CaChIzFo]   G. Casati, B.V. Chirikov, F.M. Izrailev, J. Ford, *Stochastic behavior of a quantum pendulum under a periodic perturbation*, Lect. Notes in Phys. **93** (1979) 334-351.

[CaGu]   G. Casati, I. Guarneri, *Chaos and special features of quantum systems under external perturbations*, Phys. Rev. **50** (1983) 640-643.

[Cher]   N.I. Chernov, *Ergodic and Statistical Properties of Piecewise Linear Hyperbolic Automorphisms of the 2-Torus*, to appear in J. Stat. Phys. (1992).

[ChIzSh]   B.V. Chirikov, F.M. Izrailev, D.L. Shepelyansky, *Dynamical Stochasticity in Classical and Quantum Mechanics*, Sov. Sci. Rev. C **2** (1981) 209-267; B.V. Chirikov, F.M. Izrailev, D.L. Shepelyansky, *Quantum chaos : localization vs ergodicity*, Physica D **33** (1988) 77-88.

[ChElYu]   M.D. Choi, G. Elliott, N. Yui, *Gauss polynomials and the rotation Algebra*, Inventiones. Math. (1991).

[ChDe]   V. Chulaevsky, F. Delyon, *Purely absolutely continuous spectrum for almost Mathieu operators*, (1990)

[CoFoSi]   I.P. Cornfeld, S.V. Fomin, Ya.G. Sinai, *Ergodic Theory*, Grundlerhen, Bd. **245**, Springer, Berlin, Heidelberg, New York (1982).

[Cyc] in H.L. Cycon, R.G. Froese, W. Kirsch, B. Simon, *Schrödinger Operators* , Texts and Monographs in Physics, Springer-Verlag, New York, 180 (1987).

[DiSi] E.I. Dinaburg, Ya.G. Sinai, *On the One Dimensional Schrödinger Operator with a Quasi periodic Potential*, Funct. Anal. Appl. **9** (1975) 279-289.

[Dix] J. Dixmier, *Les C*-algèbres et leurs représentations*, Paris, Gauthiers-Villars (1969).

[FiGrPr] S. Fishman, D.R. Grempel, R.E. Prange, *Chaos, Quantum Recurrences and Anderson Localization*, Phys. Rev. Lett. **49** (1982) 509-512.; D.R. Grempel, S. Fishman, R.E. Prange, *Localization in an incommensurate potential : an exactly solvable model*, Phys. Rev. Lett. **49** (1982) 833-836.

[FiHuXX] S. Fishman, N. Hurwitz, J. Phys. A (1991).

[FrSp] J. Fröhlich, T. Spencer, *Commun. Math. Phys.* **88**, 151-184 (1983).

[FrMaScSp] J. Fröhlich, F. Martinelli, E. Scoppola, T. Spencer, Commun. Math. Phys. **101** (1985) 21.

[Gua] I. Guarneri, Private Communication, Trieste (1986).

[Gut:Hou] M.C. Gutzwiller, *The semiclassical quantization of chaoti Hamiltonian systems*, NATO ASI, Les Houches, Session LII, *Chaos and Quantum Physics*, M.-J. Giannoni, A. Voros, J. Zinn-Justin Eds., North Holland, Amsterdam, London, New York, Tokyo (1991) 201-250.

[HeSj:Har1] B. Helffer, J. Sjöstrand, *Analyse semi-classique pour l'équation de Harper I avec application à l'étude de l'équation de Schrödinger avec champ magnétique*, Bulletin de la Société Mathématique de France, Tome 116, Fasc. 4, Mémoire 34 (1990).

[HeSj:Har2] B. Helffer, J. Sjöstrand, *Analyse semi-classique pour l'équation de Harper II*, Bulletin de la Société Mathématique de France, Tome 118, Fasc. 1, Mémoire 40 (1990).

[HeSj:Har3] B. Helffer, J. Sjöstrand, *Analyse semi-classique pour l'équation de Harper III*, Bulletin de la Société Mathématique de France, Tome 117, Fasc. 4, Mémoire 43 (1989).

[HeTo] E.J. Heller, S. Tomsovic, Private Communications (1991).

[HiSk] M. Hilsum, G. Skandalis, *Invariance par homotopie de la signature à coefficients dans un fibré presque plat*, Preprint (1990).

[IzSh]   F.M. Izrailev, D.L. Shepelyansky, *Quantum Resonance for a rotator in a non linear periodic field*, Teor. Mat. Fiz. **43** (1980) 417-428.

[Ker]   Ph. Kerdelhué, *Equation de Schrödinger magnétique périodique avec symétries triangulaires et hexagonales. Structure hiérarchique du spectre*, Thesis, Université de Paris-Sud, Orsay (1992).

[MeAs]   D. Mermin, N. Ashcroft, *Solid State Physics*, Saunders, Philadelphia, Tokyo (1976).

[PaChRa]   B. Pannetier, J. Chaussy, R. Rammal, *Experimental determination of the (H,T) phase diagram of a superconducting network*, J. Phys. France Lettres **44** (1983) L-853 - L-858; B. Pannetier, J. Chaussy, R. Rammal, J.-C. Villegier, *Experimental Fine Tuning of the Frustration : 2D Superconducting Network in a Magnetic Field*, Phys. Rev. Lett. **53** (1984) 1845-1848.

[Pas]   L.A. Pastur, *Spectral Properties of Disordered Systems in One Body Approximation*, Commun. Math. Phys. **75** (1980) 179.

[Ped]   G. Pedersen, *C*-algebras and their automorphism groups*, Academic Press, London, New York (1979).

[Pei]   R.E. Peierls, *Zur Theorie des Diamagnetismus von Leitungelectronen*, Z. für Phys. **80** (1933) 763-791.

[PiVo]   M. Pimsner, D. Voiculescu, *Exact sequences for K-groups and Ext groups of certain cross-product C*-algebra*, J. Operator Theory **4** (1980) 93-118.

[RaBe:Alg]   R. Rammal, J. Bellissard, *An algebraic semiclassical approach to Bloch electrons in a magnetic field*, J. Phys. France **51** (1990) 1803-1830.

[Rie]   M.A. Rieffel, *Irrational rotation C*-algebra* dans *Short Communication to the Congress of Mathematicians* (1978); M.A. Rieffel, *C*-algebras associated with irrational rotations*, Pac. J. Math. **95** (1981) 415-419.

[Sim]   B. Simon, *Almost periodic Schrödinger operators. A review* , Adv. Appl. Math. **3** (1982) 463-490.

[Sla]   I thank M. Rieffel for this information.

[Tom]   J. Tomiyama, *Topological representations of C*- algebras*, Tohoku Math. J. **14** (1962) 187-204.

[Vai]   S. Vaienti, *Ergodic Properties of the Discontinuous Sawtooth Map*, J. Stat. Phys. **67** (1992) 251.

[Wil:Cri]   M. Wilkinson, *Critical properties of electron eigenstates in incommensurate systems*, Proc. Roy. Soc. Lond. A **391** (1984) 305-350.

[Zak]   J. Zak, *Magnetic Translation Group*, Phys. Rev. A **134** (1964) 1602-1607; *Magnetic Translation Group II : Irreducible Representations*, Phys. Rev. A **134** (1964) 1607-1611.

Figure captions

Fig.1 Spectrum of Harper's model (Hofstadter's butterfly).

Fig.2 Magnetic translations and fluxes through elementary cells ; upper figure : square lattice ; lower figure : triangular lattice, from [BeKrSe].

Fig.3 Spectrum of triangular lattice with $\eta = 2\pi 0.0175$ around half flux, from [BeKrSe].

Fig.4 Spectrum of square lattice with second nearest neighbour interaction, from [BaKr].

Fig.5 Asymmetry of the central band edges for the Harper model near $\alpha = 1/3$.

Fig.6 Conical contact between bands at half flux in the Harper model.

Fig.7 Spectrum of the Hamiltonian with second nearest neighbour interaction near half flux, from [BaKr].

Fig.8 Parabolic contacts between bands at half flux in a Harper-like model, with third nearest neighbour interaction, from [BaFl].

Fig.9 Braiding of Landau sublevels in a model with second nearest neighbour interaction, from [BaKr].

Fig.10 Braiding of Dirac sublevels near half flux in a model with third nearest neighbour interaction, from [BaFl].

Fig.11 Time evolution of the kinetic energy for the standard map in the chaotic regime $K = 4$; the staight line corresponds to the classical energy and points represent quantum curves for different values of the effective Planck constant.

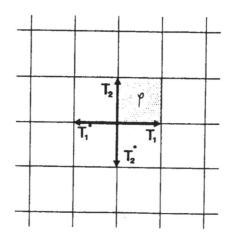

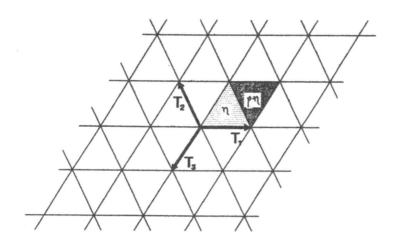

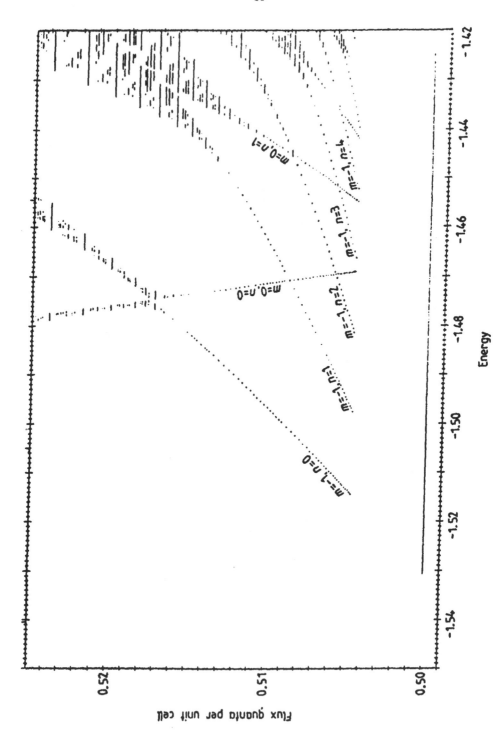

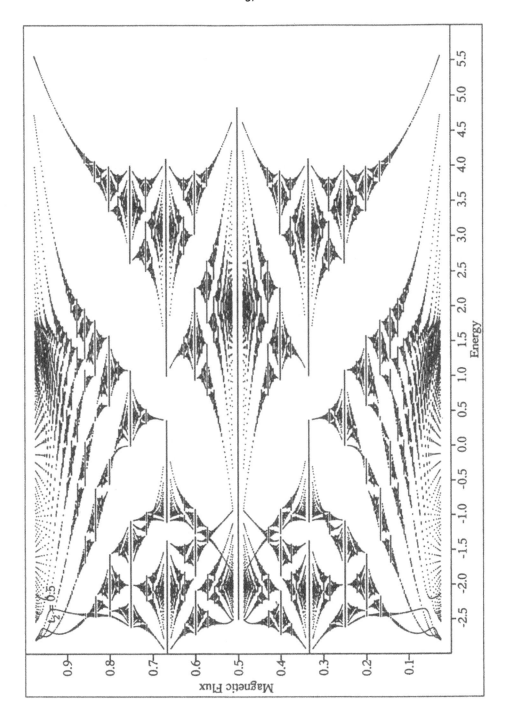

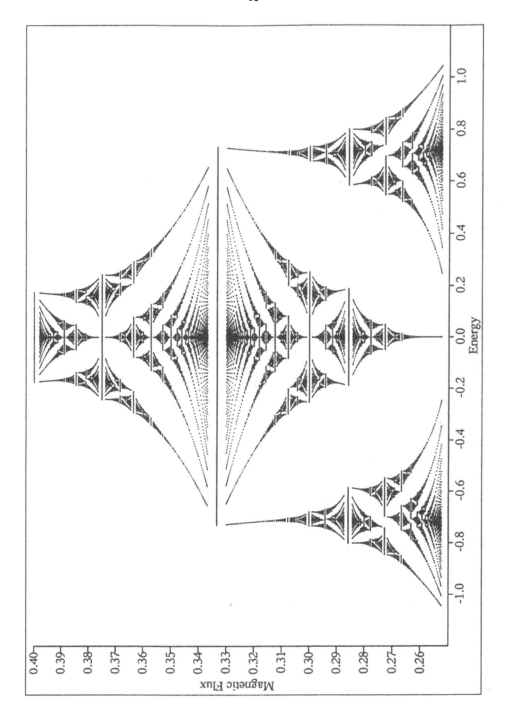

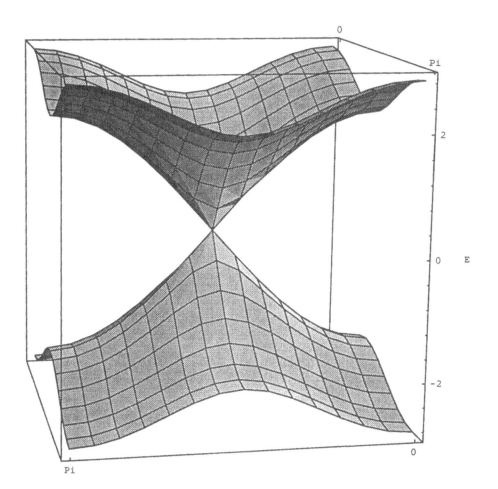

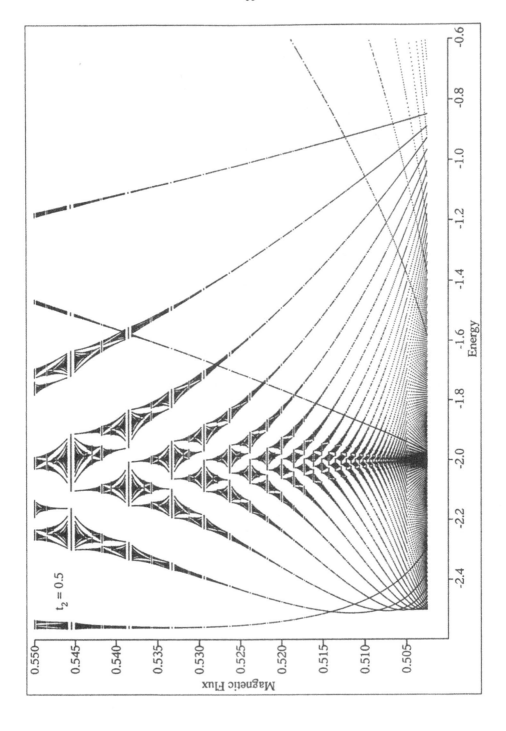

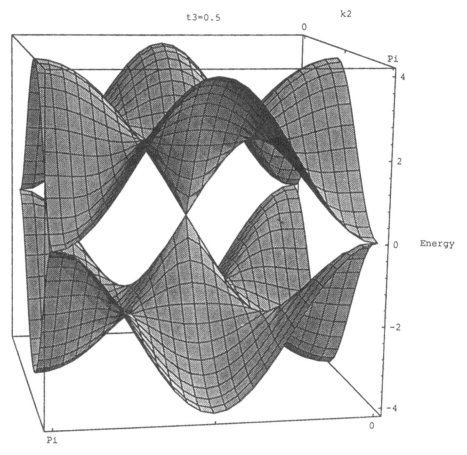

t3=0.5

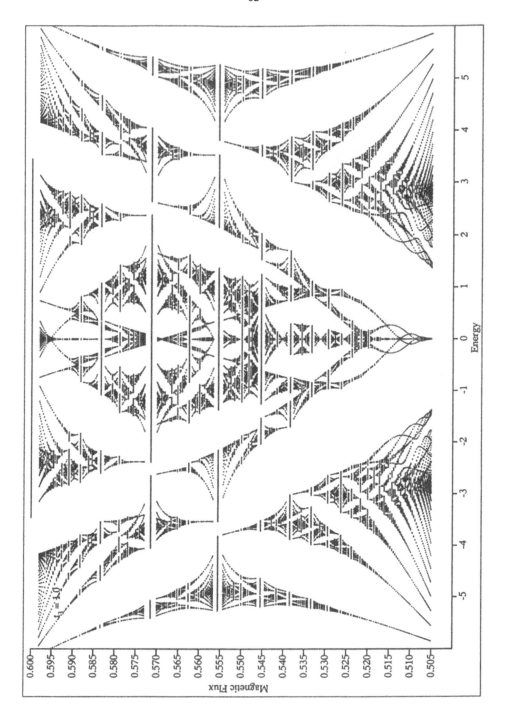

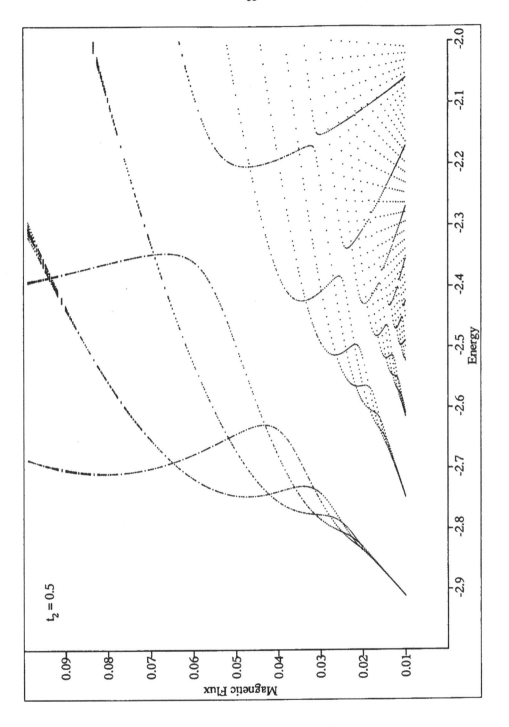

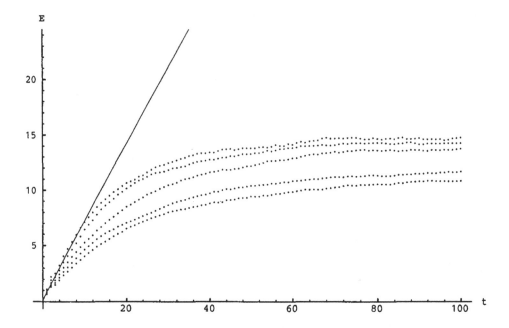

EQUIDISTRIBUTION OF PERIODIC ORBITS:
AN OVERVIEW OF CLASSICAL VS QUANTUM RESULTS

Mirko Degli Esposti, Sandro Graffi, Stefano Isola

*Dipartimento di Matematica, Università degli Studi di Bologna,
Piazza di Porta S.Donato 5, I-40127 Bologna, Italy*

1. Introduction.

Two mathematical problems, arising in classical and quantum dynamics respectively, are intimately connected to the chaotic behaviour and seem to have intriguing, though so far somewhat hidden, connections to each other.

We state first the classical problem. Let $\phi_t : M \to M$ be a time evolution on a compact manifold M where the time variable t may be discrete or continuous.

Denote by P the set of all primitive (i.e. non repeated) periodic orbits γ of it. Let $p(\gamma)$ be the (minimal) period of γ and assume that there are p.o.'s of arbitrarily large period. Denote by $\pi(x)$ the number of such γ's with period less or equal to T, i.e.

$$\pi(T) = \#\{\gamma \in P \,|\, p(\gamma) \leq T\} \tag{1.1}$$

The first part of the classical problem can therefore formulated as follows

C1 *Find the asymptotic behaviour of $\pi(T)$ as $T \to \infty$.*

To formulate the second part, given a periodic orbit γ let us define a the probability measure μ_γ in the following way: for any $f \in C(M)$, let

$$\int f \, d\mu_\gamma = \frac{1}{p(\gamma)} \int_0^{p(\gamma)} f(\phi_t(x_\gamma)) \, dt, \qquad x_\gamma \in \gamma \tag{1.2}$$

if t is a continuous variable, whereas

$$\int f \, d\mu_\gamma = \frac{1}{p(\gamma)} \sum_{t=0}^{p(\gamma)-1} f(\phi_t(x_\gamma)), \qquad x_\gamma \in \gamma \tag{1.3}$$

for discrete time evolutions.

Then the second part of the classical problem is:

C2 *Given the sequence $\{\gamma_k\}$ of all periodic orbits of ϕ_t, ordered according to non-decreasing periods, find (possibly in some suitably averaged form) all weak* limit points of the sequence $\{\mu_{\gamma_k}\}$ as $k \to \infty$.*

In other words, one would like to know how the periodic orbits become distributed in a spatial sense as their period increases.

These two questions are intimately related and have been partially solved only for a quite narrow class of dynamical systems satisfying strong hyperbolicity conditions, like the geodesic flow on a negatively curved compact manifold or, more generally, systems satisfying Smale's Axiom A [Sm]. Concerning the first question, consider for instance an hyperbolic flow restricted to its basic set $\phi_t : \Lambda \to \Lambda$ (see e.g. [PP], chap. 9, for definitions) and assume mixing. An example is the geodesic flow on a compact Riemannian manifold with sectional curvature < 0 everywhere, not necessarily constant. Then given a function $B : \Lambda \to I\!\!R$ one defines the Ruelle zeta function [Ru1]:

$$\zeta(s) = \prod_{\gamma \in P} \left[1 - \exp\left(-\int_0^{p(\gamma)} (s - B(\phi_t(x_\gamma)))dt \right) \right]^{-1} \tag{1.4}$$

and also the pressure $P(B)$ of the function B through the variational principle (see [Ru2], [PP], [Wa]):

$$P(B) = \sup_\rho \left\{ h_\rho(\phi_1) + \int_\Lambda B d\rho \right\} \tag{1.5}$$

where the sup is taken over the set of all probability measures that are invariant under the flow ϕ_t and $h_\rho(\phi_1)$ denotes the measure theoretic entropy of ρ with respect to the time one map ϕ_1 (see, e.g., [Wa]). One then shows that for B Hölder continuous the zeta function is non-zero and analytic for $\text{Re}\, s \geq P(B)$, except for a simple pole at $P(B)$. Then, considering the particular case $B = 0$, and using the above zeta function in the same way as the Riemann zeta function is used to prove the prime number theorem (that is through Wiener-Ikehara Tauberian theorem, see, e.g., [Ko]), Parry and Pollicott [PP1] proved the following 'prime orbit theorem':

$$\pi(T) \sim \frac{e^{hT}}{hT} \quad \text{as} \quad T \to \infty \tag{1.6}$$

where $h = P(0)$ is the topological entropy of the flow.

In the special case of the geodesic flow on a manifold of negative curvature one recovers an earlier result of Margulis on the distribution of lengths of closed geodesics. However for a precise understanding on how formulae of the type (1.6) can be established in different contexts we refer the reader to the following literature: [Hu], [SI1], [Mar], [B2], [PP1], [Po]. In chapter 2 we will obtain an analogous result for the linear hyperbolic diffeomorphism of the 2-torus.

We observe that far beyond the questions formulated above there is the problem of characterizing the detailed structure of the set P and in particular to obtain corrections (i.e. error terms) to the leading asymptotic behaviour (1.6). This is in general a very difficult problem belonging to an area of active mathematical research where very subtle questions are studied as testing grounds for techniques of analytic number theory.

Concerning the second question, one generically finds that the periodic orbits of a hyperbolic system not only do proliferate exponentially but their average distribution in phase space becomes so uniform that any nonperiodic trajectory can be approximated arbitrarily closely, uniformly over arbitrary long time intervals, by some periodic orbit [B2].

Consider again an hyperbolic flow restricted to its basic set Λ. Then for B Hölder continuous there exist a unique *equilibrium state* for B, i.e. an invariant probability measure μ_B which attains the supremum in the variational principle (1.5): $P(B) = h_{\mu_B}(\phi_1) + \int_\Lambda B d\mu_B$ [Ru2]. In particular, if $B = 0$ one finds the *measure of maximal entropy* μ_0 of ϕ_t. Then using some analytic properties of the pressure and again the Wiener-Ikehara Tauberian theorem one proves the following result ([Pa], [PP]):

Theorem 1.1. *Let $\phi_t : \Lambda \to \Lambda$ be a hyperbolic flow and let $B : \Lambda \to \mathbb{R}$ be a Hölder continuous function with equilibrium state μ_B. If $f \in C(\Lambda)$ then*

$$\frac{1}{Z(B,T)} \sum_{p(\gamma) \leq T} \int f \, d\mu_\gamma \cdot e^{p(\gamma) \int B \, d\mu_\gamma} \to \int_\Lambda f \, d\mu_B \qquad as \qquad T \to \infty \qquad (1.7)$$

where

$$Z(B,T) = \sum_{p(\gamma) \leq T} e^{p(\gamma) \int B \, d\mu_\gamma}$$

In particular ($Z(0,T) = \pi(T)$),

$$\frac{1}{\pi(T)} \sum_{p(\gamma) \leq T} \int f \, d\mu_\gamma \to \int_\Lambda f \, d\mu_0 \qquad as \qquad T \to \infty \qquad (1.8)$$

This result provides a weak* convergence of (weighted) averages of the sequences of measure $\{\mu_\gamma\}$ as the period of the γ's increases. One may then wonder if some stronger results can be proved under suitable assumptions. For instance, if it is possible to extract subsequences $\{\mu_{\gamma_k}\}$ of full density converging individually to the maximal entropy measure. In chapter 2 we shall prove this kind of results for the particular example of the linear diffeomorphism of the 2-torus. On the other hand, it is known that for hyperbolic systems *any* invariant probability measure can be approximated by suitable subsequences of the type $\{\mu_{\gamma_k}\}$ [HPS], so that one cannot expect in general to obtain convergence to the equilibrium state of any sequence of periodic orbits with increasing periods.

We now state the corresponding quantum problem.
Consider a compact Riemannian manifold V equipped with the standard form metric $ds^2 = g_{ij} dx^i dx^j$. Let $M = T^*V$ be the cotangent bundle of V and $\phi_t : M \to M$ the geodesic flow on it, generated by the Hamiltonian

$$H(x, \xi) = g^{ij} \xi_i \xi_j \qquad (1.9)$$

By standard quantization procedure ([LL], [La]) to the Hamiltonian (1.9) corresponds the Laplace-Beltrami operator $\Delta : C^\infty(V) \to C^\infty(V)$ given by

$$\Delta = \frac{1}{\sqrt{g}} \frac{\partial}{\partial x^i} \left(\sqrt{g} g^{ij} \frac{\partial}{\partial x^j} \right) \qquad (1.10)$$

where $g = \det(g_{ij})$. The stationary states of a quantum particle moving on V are then described by the eigenfunctions of Δ given by

$$-\hbar^2 \Delta \psi_k = \lambda_k \psi_k \qquad (1.11)$$

where the λ_k's are the eigenvalues of Δ labelled in the increasing order:

$$0 < \lambda_1 \leq \lambda_2 \leq \ldots \leq \lambda_k \leq \qquad (1.12)$$

Wee see that in this case the classical limit $\hbar \to 0$ corresponds to the large eigenvalue limit $k \to \infty$. Now, let $N(\lambda)$ be the number of λ_k's less or equal to λ, i.e.

$$N(\lambda) = \#\{\lambda_k \,|\, \lambda_k \leq \lambda\} \qquad (1.13)$$

The Weyl theorem asserts that

$$N(\lambda) = C_d \mathrm{Vol}(V)\lambda^{d/2} + R(\lambda), \quad \text{where} \quad R(\lambda) = o(\lambda^{d/2}) \qquad (1.14)$$

Here $C_d^{-1} = 2^d \pi^{d/2}\Gamma(\frac{d}{2} + 1)$ and $d = \dim V$. Then the first part of the quantum problem can be formulated as follows:
Q1 *Characterize the asymptotic behaviour of $R(\lambda)$.*

One important item in the theory of quantum chaos is that generically $R(\lambda)$ behaves as a random function so that its detailed study constitutes a very difficult mathematical problem. One way to describe this randomness is to introduce the family of measures:

$$\rho_\lambda(a,b) = \frac{1}{N(\lambda)} \#\{\lambda_k \leq \lambda, \, \lambda_k - \lambda_{k-1} \in (a,b)\} \qquad (1.15)$$

for given $0 < a < b < \infty$, and to study its weak* limit points as $\lambda \to \infty$. Such limiting distributions describe the asymptotic statistical properties of spacings between nearest eigenvalues and are expected to be completely determined by the ergodic properties of the classical geodesic flow. However, we shall not discuss further this problem here, referring the reader to the relevant literature (see, e.g., [SI2], [Sa], [UZ], [GVZ] and references therein).
    Now, let $dv(x) = \sqrt{g}dx$ be the Riemannian measure associated the the metric ds^2 and let $d\mu(x,\xi) = dv(x) \cdot d\lambda(\xi)$ be the normalized Liouville measure on $M = T^*V$. Take moreover the ψ_k's forming an othonormal basis of $L^2(V)$. Then, suppose we are given a positivity preserving quantization procedure for the observables (see e.g. [Ta]), i.e. such that to any non-negative $f \in C^\infty(M)$ it associates a (pseudo-differential) operator $\hat{f} \geq 0$ acting on $L^2(V)$. Then the correspondence

$$< \psi_k \hat{f}, \psi_k > = \int_M f(x,\xi)\, d\mu_{\psi_k}(x,\xi) \qquad (1.16)$$

defines a probability measure μ_{ψ_k} on M. Thus, the second part of the problem can be formulate exactly as problem C2 above
Q2 *Find (possibly in some suitably averaged form) the weak−∗ limit points of the sequence of measures $\{\mu_{\psi_k}\}$ as $k \to \infty$.*

Using Egorov's theorem [T] one can preliminarily show that any limit of the μ_k's must be invariant under the geodesic flow. Hence the problem reduces to investigate which invariant measure can be recovered as a classical limit of the measures μ_{ψ_k}.

This problem is of course well defined regardless of the ergodic properties of the classical flow. However the results proved so far concern mainly the two opposite situations:

1) The geodesic flow is completely integrable or quasi-integrable. In this case one is led to study how the measures μ_{ψ_k} localize on invariant lagrangian sub-manifolds. Since our main interest in this paper is to understand the possible relations between the classical equidistribution problem stated above and the present one, we shall not discuss this case here, referring the reader to the monograph of Lazutkin [La].

2) The geodesic flow is ergodic with respect to the Liouville measure (in particular is a hyperbolic flow). In this case we have the following result of Schnirelman [Schn], Zelditch [Z] and Colin de Verdiere [CdV]:

Theorem 1.2. *Assume that the geodesic flow on M is ergodic, then there exist a subsequence $\{\lambda_{k_i}\}$ of density one such that*

$$\lim_{i \to \infty} \int_M f \, d\mu_{\psi_{k_i}} = \int_M f \, d\mu \tag{1.17}$$

It is worth noticing that a preliminary step to prove (1.17) is a result on the average analogous to (1.8):

$$\frac{1}{N(\lambda)} \sum_{\lambda_k \leq \lambda} \int_M f \, d\mu_{\psi_k} \to \int_M f \, d\mu \qquad \text{as} \qquad \lambda \to \infty \tag{1.18}$$

which follows from classical symbolic calculus and the Karamata Tauberian theorem. Using the ergodicity of the classical flow and the Egorov theorem one is then able to extract a subsequence of density one converging to the r.h.s (see, e.g., [CdV]).

Thus, the above theorem asserts that almost all of the measures μ_{ψ_k} become equidistributed in the classical limit with respect to the Liouville measure. One interesting problem is then to find the conditions under which the Liouville measure is the only weak* limit point ('quantum unique ergodicity' in the language of [Sa]) so that one may avoid to take the limit on subsequences. In the next chapter we shall show that this happens in the case for the linear hyperbolic diffeomorphism of the 2-torus and its quantized version. For other results and/or conjectures in this direction see [Z1], [Z2], [RuSa].

The two problems (C) and (Q) stated above will be referred to in sequel as the *classical* and *quantum equidistribution problem*, respectively. Besides their own interest as individual mathematical problems one is also interested in their possible connections.

A first known fact is the following: consider the particular situation of a compact manifold V of dimension d and constant negative sectional curvature, say -1. Then, every conjugacy class of the fundamental group of V contains exactly one closed geodesic so that one can arrange the set of all closed geodesics as a countable family $\{\tau_k\}_{k \in \mathbb{N}}$ with non-decreasing lengths $\{\ell_k\}_{k \in \mathbb{N}}$. Moreover, there is a one-to-one correspondence between the closed geodesics τ_k of V and the primitive periodic orbits γ_k of the associated geodesic flow $\phi_t : M \to M$. In particular γ_j will have least period ℓ_k, i.e. $p(\gamma_k) = \ell_k$. Now, a remarkable relation between the two sequence of real numbers $\{\ell_k\}$ (often referred to as the *length spectrum*) and $\{\lambda_k\}$ is provided by the *Selberg trace formula* [He1]:

$$\sum_k f(\sqrt{1/4 - \lambda_k}) = \frac{\text{Vol}(V)}{4\pi} \int_{-\infty}^{+\infty} \left(-\frac{d\tilde{f}}{d\tau}\right) \frac{d\tau}{\sinh(\tau/2)} + \sum_k \sum_{n=1}^{\infty} \frac{\ell_k}{\sinh(n\ell_k/2)} \tilde{f}(n\ell_k)$$

where $f : \mathbb{R} \to \mathbb{R}$ is any C^∞ function of compact support and $\tilde{f}$ is its Fourier transform. A great deal of mathematical work has been made around this formula and several applications and generalisations have been proposed (see [He2], [He3], [BalVor] and references therein).

On the other side, very little is known about the possible connections between the two sequences of measures $\{\mu_{\gamma_k}\}$ and $\{\mu_{\psi_k}\}$ defined respectively in (1.2) and (1.16). One remark is the following. It is known that for a geodesic flow on a negatively curved manifold, the Liouville measure coincides with the measure of maximal entropy when the curvature is constant (see, e.g., [K]). Thus, one may argue that in this case there should exist some direct relation between (1.8) and Theorem 1.2 (in the next chapter we will examine a discrete time dynamical system where these two measures coincides as well and the connection can be established explicitly). On the other hand, for the more general case of variable curvature the problem seems much more involved.

2. An example: the hyperbolic linear automorphism of the torus.

We now consider the discrete time dynamical system $T : M \to M$ where M is the 2−torus $T^2 = \mathbb{R}^2/Z^2$ (points on T^2 are denoted by $x = (p,q) \in [0,1] \times [0,1]$) and T is the hyperbolic automorphism of T^2 generated by the matrix

$$A = \begin{pmatrix} a & b \\ c & d \end{pmatrix} \tag{2.1}$$

such that $(a,b,c,d) \in Z$, $ad - bc = 1$ and $|a + d| > 2$. The Lebesgue measure μ is invariant because $\det A = 1$. Moreover, the condition $|TrA| > 2$ makes this dynamical system an Anosov one and hence, in particular, ergodic and mixing with respect to μ.

Now, an orthonormal basis in $L^2(T^2, d\mu)$ is given by the set

$$\{T(n) = e^{2\pi i <n,x>} | n \in Z^2\} \tag{2.2}$$

and A acts on points $x = (q,p)$ and on suitably smooth functions $f(x)$ on M respectively as:

$$Ax = ((aq + bp)(\text{mod } 1), (cq + dp)(\text{mod } 1))$$

$$f(Ax) = \sum_{n \in Z^2} f_n T(A^t n) \tag{2.3}$$

where A^t is the transposed matrix of A (notice that we have used the same symbol A to denote the matrix A and the map T; this will be repeatedly done in what follows without fear of confusion).

Consider then in $L^2(T^2, d\mu)$ the unitary *Koopman operator* $\mathcal{U}_A$ defined by

$$(\mathcal{U}_A f)(x) = f(Ax) \tag{2.4}$$

and recall (see, e.g., [A.A]) that T is ergodic iff 1 is a simple eigenvalue of $\mathcal{U}_A$. Otherwise stated, if there is $h \in L^2(T^2, d\mu)$ such that $\mathcal{U}_A h = h$ then h is constant μ-almost everywhere. Moreover, it is mixing iff, for any pair $f, g \in L^2(T^2, d\mu)$,

$$\lim_{k \to \infty} <\mathcal{U}_A^k f, g> = <f, 1><1, g> \tag{2.5}$$

This property makes $\sigma(\mathcal{U}_A)$ continuous on the unit circle, but for the eigenvalue 1.

Now, if A is a continuous map of a compact metric space X then $h_{top}(A)$, the topological entropy, satisfies the restricted variational principle:

$$h_{top}(A) = \sup_{\mu \in \mathcal{M}_A(X)} h_\mu(A) \tag{2.6}$$

where $\mathcal{M}_A(X)$ the set of the probability measures on the Borel σ-algebra of X which are A-invariant and $h_\mu(A)$ is the measure theoretic entropy (see [M], [AY]). Moreover, A is *intrinsically ergodic* if there exist a unique $\mu \in \mathcal{M}_A(X)$ such that $h_{top}(A) = h_\mu(A)$. In this case μ is called the intrinsic measure of A, or maximal entropy measure of A. Thus, for a linear automorphism of the torus the Haar measure, i.e. the Lebesgue one, is actually an intrinsic measure. We have the

Theorem 2.1 (Sinai) *Let $A : T^d \to T^d$ be a linear automorphism with eigenvalues $\lambda_1, \ldots \lambda_d$. Let μ be the Haar measure of T^d. Then*

$$h_{top}(A) = h_\mu(A) = \sum_{|\lambda_i| \geq 1} \log |\lambda_i| \tag{2.7}$$

Proof. See e.g. [A.A]. Notice that this result can be easily obtained also from Pesin's formula ([M], p.265).

Consider again the general situation of a continuous map A of a compact metric space X. We say that $x \in X$ is a *periodic point* of A, of period n, if it is a fixed point of A^n, i.e. $A^n x = x$. We denote by Fix_n the set of such points. It is easy to see that for linear automorphism of the torus the set of periodic points of A is dense in T^2, because it coincides with the subset of T^2 formed by all points having rational coordinates (see below). More generally we have the

Theorem 2.2 (Bowen-Sinai) *Every topologically mixing hyperbolic homeomorphism $A : X \to X$ is intrinsically ergodic. If μ denotes its intrinsic measure then for every continuous map $f : X \to \mathbb{R}$:*

$$\int_X f d\mu = \lim_{n \to \infty} \frac{1}{\#Fix_n} \sum_{x \in Fix_n} f(x) \tag{2.8}$$

and A has topological entropy

$$h_{top} = \lim_{n \to \infty} \frac{1}{n} \log \#Fix_n \tag{2.9}$$

Proof. See [M], p.254.

For the linear automorphism of the torus we have the additional result:

Proposition 2.1. *Let λ be the eigenvalue of A whose modulus is larger than 1. Then*

$$\#Fix_n = \lambda^n + \lambda^{-n} - 2 \tag{2.10}$$

Proof. We sketch the idea of the proof referring to [I] for more details. Let k be the trace of A, then $|k| > 2$. Consider the numbers

$$u_n = \frac{\lambda^n - \lambda^{-n}}{2D} \tag{2.11}$$

where $D = \sqrt{\frac{k^2}{4} - 1}$. These numbers satisfy the recursion

$$u_0 = 0, u_1 = 1 \quad \text{and} \quad u_n = k u_{n-1} - u_{n-2}, \quad \text{for } n > 1$$

and it can be easily checked by induction that

$$A^n = \begin{pmatrix} a u_n - u_{n-1} & b u_n \\ c u_n & d u_n - u_{n-1} \end{pmatrix} \tag{2.12}$$

By virtue of this formula one can easily realize that for any $n > 0$ the set Fix_n constitutes a regular lattice on the torus and by a simple geometrical argument $\#Fix_n$ can be computed as the inverse of the area of an elementary cell in the above lattice (see [I]), thus giving (2.10). Q.E.D.

Remark. Notice that from (2.9) and (2.10) one immediately recovers (2.7) for this particular case: $h_\mu(A) = h_{top}(A) = \log \lambda$.

We now deduce some further consequences of (2.9) and (2.10). Denote again by P the set of all primitive periodic orbits (prime cycles) of A and by $\pi'(n)$ and $\pi(x)$ the number of them whose period is n and less or equal to x respectively, i.e.

$$\pi'(n) = \#\{\gamma \in P | p(\gamma) = n\} \qquad \pi(x) = \#\{\gamma \in P | p(\gamma) \leq x\} \tag{2.13}$$

Proposition 2.2.

$$\pi'(n) \sim \frac{\lambda^n}{n} \tag{2.14}$$

and

$$\pi(x) \sim \frac{\lambda}{\lambda - 1} \cdot \frac{\lambda^x}{x} \tag{2.15}$$

where $f(t) \sim g(t)$ means that $f(t)/g(t) \to 1$ when $t \to \infty$.

Proof. We make use of the zeta function

$$\zeta(z) = \exp \sum_{n=1}^{\infty} \frac{z^n}{n} \sum_{x \in Fix_n} 1 \tag{2.16}$$

The strategy is to gain insight on the distribution of closed orbits out of the mero-morphy domain of $\zeta(z)$. In our case, by Proposition 2.1, $\zeta(z)$ has the simple form

$$\zeta(z) = \frac{(1-z)^2}{(1-\lambda z)(1-z/\lambda)} \tag{2.17}$$

so that

$$\frac{\zeta'(z)}{\zeta(z)} = \frac{\lambda}{1-\lambda z} + g(z) \tag{2.18}$$

where $g(z)$ is analytic in $\{z||z| < e^\epsilon/\lambda\}$ for some $\epsilon > 0$. The rest of the proof proceeds exactly in the same way as in the proof for subshifts of finite type ([P.P], p.100). Q.E.D.

Remark. Proposition 2.2 is the analogue of the *prime orbit theorem* proved in the context of Axiom A flows (see [P.P]).

We now prove that closed orbits exhibit a regularity in a spatial sense. In particular, we show that they are equidistributed on the average with respect to the Lebesgue measure μ.

Let μ_γ be the measure defined by

$$\mu_\gamma = \frac{1}{p(\gamma)} \sum_{k=0}^{p(\gamma)-1} \delta_{A^k(x)} \qquad x \in \gamma \tag{2.19}$$

and set $\int_\gamma f = \int_{T^2} f d\mu_\gamma$. Then we have the

Proposition 2.3. *For every continuous map* $f : T^2 \to \mathbb{R}$:

$$\frac{1}{\pi'(n)} \sum_{p(\gamma)=n} \int_\gamma f \longrightarrow \int_{T^2} f d\mu \quad as \quad n \to \infty \tag{2.20}$$

Proof. We first write the number of fixed points of period n in the form

$$\#Fix_n = \sum_{l|n} l \cdot \pi'(l) \tag{2.21}$$

where $l|n$ means that l divides n. Form Proposition 2.1 we then have

$$\sum_{l|n} l \cdot \pi'(l) = \lambda^n + \lambda^{-n} - 2 \quad . \tag{2.22}$$

On the other hand Theorem 2.2 yields

$$\lim_{n\to\infty} \frac{\sum_{l|n} l \cdot \sum_{p(\gamma)=l} \int_\gamma f}{\sum_{l|n} l \cdot \pi'(l)} = \int_{T^2} f d\mu \tag{2.23}$$

Hence,

$$\sum_{l|n} l \cdot \sum_{p(\gamma)=l} \int_\gamma f \sim (\lambda^n + \lambda^{-n} - 2) \int_{T^2} f d\mu, \quad \text{as} \quad n \to \infty \qquad (2.24)$$

Let $m = \max\{l \in Z| \ l|n, \ l < n\}$. If n is prime then $m = 1$, otherwise $m > 1$. One finds immediately that

$$\sum_{p(\gamma)=n} \int_\gamma f = \frac{1}{n}\left(\sum_{l|n} l \cdot \sum_{p(\gamma)=l} \int_\gamma f - \sum_{l|n, \ l \le m} l \cdot \sum_{p(\gamma)=l} \int_\gamma f\right)$$

from which we obtain

$$\sum_{p(\gamma)=n} \int_\gamma f \sim \frac{\lambda^n}{n}(1 - C_{n,m}) \int_{T^2} f d\mu, \quad \text{as} \quad n \to \infty \qquad (2.25)$$

where $C_{n,m}$ is of order at most λ^{m-n}. To conclude the proof it suffices to divide (2.25) by $\pi'(n)$ and apply Proposition 2.2. Q.E.D.

Proposition 2.3 yields a result of uniform distribution on the average. We now study the behaviour of measures μ_γ supported on single closed orbits and we prove that they converge in measure to μ.

Proposition 2.4. *For any $\epsilon > 0$ and for any continuous function f*

$$\lim_{n\to\infty} \left(\frac{\#\{\gamma|p(\gamma) = n, \ |\int_\gamma f - \int_{T^2} f d\mu| > \epsilon\}}{\pi'(n)}\right) = 0 \qquad (2.26)$$

Proof. For the sake of simplicity we shall use the notation $\int_{T^2} f d\mu = \bar{f}$. For any $k \in Z_+$ set

$$m_k f(x) = \frac{1}{k}\sum_{l=0}^{k-1} f(A^l x) \qquad (2.27)$$

Now, given $\delta > 0$, introduce the sets

$$S_{\delta,n} = \{\gamma|p(\gamma) = n, \ \left|\int_\gamma f - \bar{f}\right| \le \sqrt{\delta}\}$$
$$R_{\delta,n,k} = \{\gamma|p(\gamma) = n, \ \int_\gamma |m_k f - \bar{f}| \le \sqrt{\delta}\} \qquad (2.28)$$

We have

$$\left|\int_\gamma f - \bar{f}\right| \le \left|\int_\gamma (f - m_k f)\right| + \int_\gamma |m_k f - \bar{f}|$$

On the other hand it is obvious that $\int_\gamma (f - m_k f) = 0$, for any $k \in Z_+$; and thus

$$S_{\delta,n} \supset R_{\delta,n,k} \Longrightarrow R_{\delta,n,k}^c \supset S_{\delta,n}^c \quad \text{for any} \quad \delta > 0, \ k \in Z_+ \qquad (2.29)$$

The ergodicity of A implies that, for μ-almost every $x \in T^2$,

$$\lim_{k\to\infty} m_k f(x) = \bar{f} \qquad (2.30)$$

Hence, for any continuous f, by the Lebesgue dominated convergence theorem we can find a $k_0 > 0$ such that

$$\int_{T^2} |m_{k_0} f(x) - \bar{f}| d\mu \leq \frac{\delta}{2} \tag{2.31}$$

Set

$$\int_{T^2} f d\mu_n := \frac{1}{\pi'(n)} \sum_{p(\gamma)=n} \int_\gamma f$$

Then Proposition 2.3 entails that $d\mu_n$ converges vaguely to the Lebesgue measure. This implies the existence of $n_0(\delta) > 0$ such that for any $n > n_0$

$$\int_{T^2} |m_{k_0} f(x) - \bar{f}| d\mu_n \leq \delta \tag{2.32}$$

Hence by the Chebychev inequality we obtain

$$\frac{\#\{\gamma | p(\gamma) = n, \ \int_\gamma |m_{k_0} f - \bar{f}| \geq \sqrt{\delta}\}}{\pi'(n)} \leq \sqrt{\delta} \tag{2.33}$$

and therefore, by the second of (2.28):

$$\frac{\#R^c_{\delta,n,k_0}}{\pi'(n)} \leq \sqrt{\delta}, \qquad n > n_0 \tag{2.34}$$

Hence, from (2.29), we find

$$\frac{\#S^c_{\delta,n}}{\pi'(n)} \leq \sqrt{\delta}, \qquad n > n_0 \tag{2.35}$$

and the assertion follows by taking $\delta \to 0$ and consequently $n_0(\delta) \to \infty$. Q.E.D.

2.1 Koopman operator and periodic orbits on invariant lattices.

Consider any point on T^2 having coordinates $(r/N, r'/N)$, with $r, r', N \in \mathbb{N}$ and $0 \leq r, r' < N$. There are exactly N^2 points of this type and they belong to the $N \times N$ subgroup of T^2 given by:

$$L_N = \{(q, p) \in T^2 | Nq, Np \in Z\} \tag{2.36}$$

It is immediate to realize that L_N is invariant under the action of A, so that any point in L_N is periodic with period $\leq N^2$, the origin being the only fixed point of A. Of course, any point $x \in Fix_n$ belongs to a periodic orbit whose period divides n. This means that L_N splits into periodic orbits (which in general may have different periods) of A.

Let μ_N be the normalized atomic measure supported on L_N. We now characterize the spectrum of $\mathcal{U}_A$ when acting on $L^2(T^2, d\mu_N)$. Let M_N be the number of distinct periodic orbits of A^t which live on $L_N \setminus \{0, 0\}$. This number is the same as that corresponding to A (see [I]). Let $\gamma \subset L_N$ be any one of such orbits with period

$p(\gamma)$ and $x = (r_1/N, r_2/N) \in \gamma$. Then, associated to each orbit γ there are $p(\gamma)$ linearly independent vectors in $\mathcal{C}^{N^2}$ given by:

$$f_l(k) = \sum_{s=0}^{p(\gamma)-1} \lambda_l^{-s} \exp \frac{2\pi i}{N} < (A^t)^s r, k > \qquad r = (r_1, r_2), \ k \in Z_N^2 \qquad (2.37)$$

where $\lambda_l = e^{-2\pi i l/p(\gamma)}$ and $l = 0, \ldots, p(\gamma) - 1$, and they satisfy

$$\mathcal{U}_A f_l(k) = \lambda_l f_l(k) \qquad (2.38)$$

Thus, there are $N^2 - 1$ eigenvectors of $\mathcal{U}_A$ of the form (2.37) which, together with the constant function 1, provide a canonical basis of $L^2(T^2, d\mu_N)$. Among them, there are exactly M_N non-constant functions which are invariant. This is in account of the fact the dynamical system (T^2, A, μ_N) is not ergodic: the invariant measure μ_N obviously admits a decomposition into invariant ergodic measures of the type (2.19).

The case of N prime.

We now specialize now to the lattices L_N with N prime. In this case a very precise characterization of the structure of the periodic orbits is possible (for which we refer to [D.G.I]) and moreover strong results on their equidistribution properties can be proved.

First, it has been shown by Percival and Vivaldi [P.V] that all the periodic orbits living in L_N with N prime have the same period, i.e. $p(\gamma) = p(N)$ for any $\gamma \subset L_N \setminus \{0, 0\}$. The relation among $p(N)$, M_N and N is then:

$$p(N) \cdot M_N = N^2 - 1 \qquad (2.39)$$

Thus, from the above argument we then have the following result on the spectrum of the Koopman operator:

Proposition 2.5. *Let N be a prime number and let $p(N)$ be the period of the cycles living on $L_N \setminus \{0, 0\}$. Then $\sigma(\mathcal{U}_A)$ is given by the eigenvalues*

$$\lambda_l = e^{2\pi i l/p(N)} \qquad l = 0, 1, \ldots, p(N) - 1 \qquad (2.40)$$

To each λ_l is associated an eigenspace E_l of non constant functions to which all the periodic orbits of $L_N \setminus \{0, 0\}$ contribute. Accordingly, the following decomposition holds:

$$L^2(T^2, d\mu_N) = \widehat{1} \bigoplus \left(\bigoplus_{l=0}^{p(N)-1} E_l \right) \qquad (2.41)$$

where $\widehat{1}$ is the one-dimensional subspace spanned by the function 1 and $dim(E_l) = M_N \ \forall \ l = 0, \ldots, p(N) - 1$.

We now turn to the equidistribution results. We first prove a sharpening of Proposition 2.4:

Proposition 2.6. *For any $f \in C(T^2)$ and any $\epsilon > 0$ there is N_0 such that, if $N \geq N_0$ is prime:*

$$\frac{\#\{\gamma \subset L_N \setminus \{0,0\}, \ |\int_\gamma f - \bar{f}| \geq \sqrt{\epsilon}\}}{\#\{\gamma | \gamma \subset L_N \setminus \{0,0\}\}} \leq \sqrt{\epsilon} \qquad (2.42)$$

Proof. It is easy to realize that, up to the additive correction vanishing as N^{-2} for $N \to \infty$ (arising from the fixed point at the origin), the following identity holds true:

$$\frac{1}{M_N} \sum_{j=1}^{M_N} \int_{\gamma_j} f = \int_{T^2} f d\mu_N \qquad (2.43)$$

Then the assertion follows by the same argument as in Proposition 2.4 where now

$$S_{\epsilon,n} := \{\gamma | \gamma \subset L_N \setminus \{0,0\}, \ \left| \int_\gamma f - \bar{f} \right| \leq \sqrt{\epsilon}\}$$

$$R_{\epsilon,n,k} := \{\gamma | \gamma \subset L_N \setminus \{0,0\}, \ \int_\gamma |m_k f - \bar{f}| \leq \sqrt{\epsilon}\}$$

Q.E.D.

Now, under slightly more restrictive assumptions on the sequence of primes and using the number theoretic techniques collected in the Appendix, we are able to prove the equidistribution of *all* periodic orbit sequences (living on prime lattices), with explicit estimates of the speed of convergence. The key fact is that since Z_N becomes a finite field, this restriction amounts to operate with a (mod N) arithmetics.

Recall that the prime number N is *splitting* with respect to the characteristic polynomial of A if there exist $n \in Z_N$ such that $k^2 - 4 = n^2 (\mathrm{mod}\, N)$ (where k denotes the trace of A) and is otherwise *inert* (see, e.g. [H]). It can be shown (see [D.G.I]) that if N is splitting then $p(N) = (N-1)/m$ for some $m \in \mathbb{N}$ (so that there are exactly $m(N+1)$ periodic orbits living in L_N), whereas if N is inert $p(N) = (N+1)/m$.

Finally, let $A(T^2)$ be the Banach space of all functions $f : T^2 \to \mathbb{C}$ such that

$$\|f\|_A = \sum_{n \in Z^2} |f_n| < \infty \qquad (2.44)$$

We now prove the following stronger result:

Theorem 2.3. *Let $N \in \Gamma$, Γ being any increasing sequence of primes such that $N/p(N) < C$ for some C independent of N. Set:*

$$P_N = \{\gamma \subset L_N \setminus \{0,0\} \mid \gamma \text{ periodic orbit of } A\}, \qquad M_N = \#P_N.$$

Then, given any $f \in A(T^2)$, any sequence $\{\gamma_{j(N)}\}_{N \in \Gamma}$ such that $\gamma_{j(N)} \in P_N$, $j(N) \in \{1, \ldots, M_N\}$ we have

$$\lim_{N \to \infty} \left| \int_{\gamma_{j(N)}} f - \int_{T^2} f \, d\mu \right| = 0 \qquad (2.45)$$

Moreover, if $f \in C^\infty(T^2)$ there is $C > 0$ such that for N large enough:

$$\left| \int_{\gamma_j(N)} f - \int_{T^2} f \, d\mu \right| < \frac{C}{\sqrt{N}} \|f\|_{A(T^2)} \tag{2.46}$$

Remark. The condition $N \in \Gamma$ is equivalent to require m bounded with respect to N. On the other hand, the existence of at least a sequence of primes which satisfies this condition (actually the fact that almost any sequence of primes does it) is a consequence of the Artin conjecture, whose failure would imply the falsity of the generalized Riemann hypothesis (see e.g. [RM]). Detailed heuristic and numerical investigations on the behaviour of $p(N)/N$ supporting the above genericity can be found in [K2] and [B.V].

Proof. We shall give the argument for the case of N splitting. This is equivalent to the splitting of the characteristic polynomial over Z_N. In particular (see the Appendix), if we let $D = \sqrt{\frac{k^2}{4} - 1} \in Z_N$, then $v = (1, D)$, $w = (1, -D) \in Z_N^2$ are the eigenvectors of A acting on Z_N^2, corresponding to the eigenvalues $\lambda_N = k/2 + D$ and $\lambda_N^{-1} = k/2 - D$, respectively. Now, the key point is that the integral of an arbitrary character $e^{i2\pi<n,x>}$ over a periodic orbit $\gamma \subset L_N$ can be written as a Kloosterman sum restricted to a cyclic subgroup of Z_N^* of order $(N-1)/m$. Indeed, for any $\gamma \subset L_N$ and $x \in \gamma$ set $Nx = \alpha v + \beta w$ where $\alpha(x), \beta(x) \in Z_N$. Then,

$$\int_{\gamma \subset L_N} e^{2\pi i<n,x>} = \frac{1}{p(N)} \sum_{s=0}^{p(N)-1} e^{2\pi i<n,A^s x>} =$$

$$\frac{1}{p(N)} \sum_{s=0}^{p(N)-1} e^{\frac{2\pi i}{N}<n,\alpha v \lambda_N^s + \beta w \lambda_N^{-s})>} = \frac{1}{p(N)} \sum_{s=0}^{p(N)-1} e^{\frac{2\pi i}{N}(\alpha<n,v>\lambda_N^s + \beta<n,w>\lambda_N^{-s})}$$

$$\frac{1}{mp(N)} \sum_{\xi \in Z_N^*} e^{\frac{2\pi i}{N}(a\xi^m + b\xi^{-m})} = \frac{1}{mp(N)} \sum_{j=0}^{m-1} \sum_{\xi \in Z_N^*} \chi_j(\xi) e^{\frac{2\pi i}{N}(a\xi + b\xi^{-1})}$$

$$\tag{2.47}$$

where $a = \alpha < n, v >$, $b = \beta < n, w > \in Z_N$ and the relation $p(N) = (N-1)/m$ has been used. The functions $\chi_j : j = 0, \ldots, m-1$ are the m distinct multiplicative characters of order m of Z_N (see the Appendix). We can now apply the estimate (A.18) and obtain, for any $n \neq (0,0) \pmod N$:

$$\left| \int_{\gamma \subset L_N} e^{2\pi i<n,x>} \right| \leq \frac{K}{\sqrt{N}} \tag{2.48}$$

where $K > 0$ is independent of n and of the particular orbit $\gamma \subset L_N$ and is uniformly bounded in N because of the boundedness of m. On the other hand, a trivial computation shows that, if $n = (0,0) \pmod N$ then

$$\int_{\gamma \subset L_N} e^{2\pi i<n,x>} = 1 \tag{2.49}$$

Therefore if $f = \sum_{n \in Z^2} f_n e^{2\pi i <n,x>}$ we can write

$$\int_{\gamma_j(N)} f - \int_{T^2} f \, d\mu = \sum_{\substack{n \neq (0,0) (\text{mod } N)}} f_n \int_{\gamma_j(N)} e^{2\pi i <n,x>} + \sum_{k \neq (0,0)} f_{kN} \qquad (2.50)$$

Now, if $f \in A(T^2)$ the second term of the r.h.s. vanishes as $N \to \infty$ and the first term also vanishes by the uniform estimate (2.48) on the Kloosterman sums. If moreover $f \in C^\infty(T^2)$ the second term vanishes at least as N^{-1} as $N \to \infty$ and therefore we can conclude that there exists $C > 0$ such that if N is large enough:

$$\left| \int_{\gamma_j(N)} f - \int_{T^2} f \, d\mu \right| \leq \frac{C}{\sqrt{N}} \|f\|_{A(T^2)} \qquad (2.51)$$

Finally, the case N inert can be treated in a similar way by using the techniques of [PV] and the generalized Kloosterman sums over arbitrary finite fields (see [Ka1]). Q.E.D.

2.2. The quantum equidistribution problem. Wigner function and classical limit.

We now consider the quantum dynamical system $(\mathcal{H}, \mathcal{A}, V_A)$ obtained by the canonical quantization of the former one. We now limit ourselves to recall the main ingredients referring the reader to [DGI] for the detailed construction.

1) The Hilbert space is $\mathcal{H} = \mathcal{H}_N = L^2(S^1, \nu_N)$ where $h = 1/N$ and ν_N is the normalized atomic measure given by ($e^{2\pi i q} \in S^1$):

$$\nu_N(q) = \frac{1}{N} \sum_{l=0}^{N-1} \delta(q - \frac{l}{N}) \qquad (2.52)$$

Hence $\mathcal{H}$ has dimension N.

2) The algebra $\mathcal{A}$ is the *-algebra of the observables on $\mathcal{H}$ generated by the quantization of the classical functions on the torus in the following way: let $\hat{T}(n)$, $n \in Z^2$ be the canonical quantization of the basic observables $T(n)$. This is based on the classification of the irreducible representations of the discrete Heisenberg group $H_1(Z)$, in complete analogy with the well known construction of the Schrödinger representation out of the Heisenberg group $H_n(I\!R^n)$. The quantization of any $f \in A(T^2)$ is then given by

$$\hat{f} = \sum_{n \in Z^2} f_n \hat{T}(n) \qquad (2.53)$$

3) The unitary bijection V_A is the quantum propagator, i.e. the quantization of the action of the symplectomorphism A on the classical observables. This means that if $f \to \hat{f}$ as above then $f(Ax) \to V_A \hat{f} V_A^{-1}$. Therefore the quantum discrete dynamics is defined as

$$\hat{f} \to V_A^k \hat{f} V_A^{-k}, \qquad k \in Z \qquad (2.54)$$

Moreover, if N is a prime number, one easily finds that $V_A^p(N) = \text{Id}$ so that the quantum dynamics is periodic with period given by the classical periodic orbits living on the lattice L_N.

4) We denote by $e^{2\pi i \lambda_n^{(N)}}$ and $\phi_n^{(N)}$, $n = 0, \ldots N-1$, the (repeated) eigenvalues of V_A and the corresponding (orthonormal) eigenvectors, respectively.

The quantum equidistribution problem can now be stated as follows: given any pair of eigenvectors $\phi_k^{(N)}, \phi_l^{(N)}$ belonging to the same eigenspace, define a distribution $d\Omega_N(\phi_k, \phi_l)$ on the phase space T^2 by

$$\int_{T^2} f \, d\Omega_N(\phi_k^{(N)}, \phi_l^{(N)}) = < \phi_k^{(N)}, \hat{f}\phi_l^{(N)} > \tag{2.55}$$

Then, we want to know what are the weak* limit points of such distributions when $N \to \infty$. The main result of [DGI] is the following

Theorem 2.4. *Let Γ be an increasing sequence of primes as in Theorem 2.3. Then, for any sequence of eigenvectors $\{\phi_n^{(N)}\}_{N \in \Gamma}$ and $f \in A(T^2)$ we have*

$$\lim_{N \to \infty} \int_{T^2} f \, d\Omega_N(\phi_n^{(N)}) = \int_{T^2} f \, d\mu \tag{2.56}$$

and, if f is smooth enough, say $f \in C^\infty(T^2)$, the limit is attained with speed given by:

$$\left| \int_{T^2} f \, d\Omega_N(\phi^{(N)}) - \int_{T^2} f \, d\mu \right| \le \frac{C\|f\|_A}{\sqrt{N}} \tag{2.57}$$

Moreover, for any pair of sequences $\{\phi_k^{(N)}\}_{N \in \Gamma}$, $\{\phi_l^{(N)}\}_{N \in \Gamma}$ of eigenvectors belonging to the same eigenspace but however distinct we have:

$$\lim_{N \to \infty} \int_{T^2} f \, d\Omega_N(\phi_k^{(N)}, \phi_l^{(N)}) = 0 \tag{2.58}$$

and the limit is attained with the same speed as in (2.57).

Remark. This result is stronger than the other results of this kind mentioned in chapter 1 in that it holds *for any* sequence of eigenvectors ('quantum unique ergodicity' in Sarnak's language [Sa]).

In [DGI] two independent proofs of the above result have been constructed. Here we shall just recall the main ideas of the proof using the Wigner function, because it also provides some insight into the question raised in the previous chapter about the relationships beetween the distributions $d\Omega_N$ above and the measures $d\mu_\gamma$ (see (2.19)).

We then define the Wigner transform, which is a map from pairs of functions in the Hilbert space into the phase-space functions, in the following way:

Let $\phi, \psi \in L^2(Z_N, \nu_N)$, $(q, p) \in T^2$. Then their *discrete Wigner transform* is the function defined as

$$W(\phi, \psi)(q, p) = \sum_{n_1, n_2 \in Z_N} < \phi, \hat{T}(n)\psi > e^{-2\pi i(n_1 q + n_2 p)} \tag{2.59}$$

Since the Wigner functions have to be integrated against the measures ν_N on S^1 or μ_N on T^2, we are interested only in the values they assume on the lattice L_N, and therefore we use the notation:

$$W(\phi,\psi)(s,r) = W(\phi,\psi)(\frac{s}{N},\frac{r}{N}) \qquad (2.60)$$

$\forall s,r \in Z_N$. Now, using the inverse of the discrete Fourier transform we get

$$W(\phi,\psi)(s,r) = \sum_{k\in Z_N} e^{-\frac{2\pi i k r}{N}} \phi(s-\frac{k}{2})\bar{\psi}(s+\frac{k}{2})$$
$$= N \int e^{-2\pi i N 2pq'} \phi(q-q')\bar{\psi}(q+q')d\nu_N(q') \qquad (2.61)$$

In particular for $\psi = \phi$ we obtain the formula for the *Wigner distribution*

$$W_\phi(s,r) = W(\phi,\phi)(s,r) = \sum_{k\in Z_N} e^{-\frac{2\pi i 2 k m_2}{N}} \phi(s-k)\bar{\phi}(r+k) \qquad (2.62)$$

We have the

Proposition 2.7. *The following representation holds:*

$$\int_{T^2} f d\Omega_N(\phi_k^{(N)},\phi_l^{(N)}) = \int_{T^2} fW(\phi_k^{(N)},\phi_l^{(N)})d\mu_N \qquad (2.63)$$

Proof. By (2.53), (2.55) and (2.59) we have

$$\int_{T^2} f d\Omega_N(\phi_k^{(N)},\phi_l^{(N)}) = <\phi_k^{(N)},\hat{f}\phi_l^{(N)}>$$
$$= \sum_{n\in Z_N^2} f_n <\phi_k^{(N)},\hat{T}(n)\phi_l^{(N)}>$$
$$= \sum_{n\in Z_N^2} f_n \int_{T^2} e^{2\pi i <n,x>} W(\phi_k^{(N)},\phi_l^{(N)})(x)d\mu_N(x)$$
$$= \int_{T^2} f(x)W(\phi_k^{(N)},\phi_l^{(N)})(x)d\mu_N(x)$$

Q.E.D.

Thus, one is reduced to investigate the behaviour of the Wigner functions in the classical limit.

A first remarkable property is that the Wigner functions allow us to define a set of eigenvectors for the adjoint of Koopman operator acting on $L^2(T^2, d\mu_N)$ (cfr. Proposition (2.6)):

Proposition 2.8. *Let $\{W_{k,l}\}$ be the family of N^2 functions on the torus given by*

$$W_{k,l}(x) = W(\phi_k^{(N)},\phi_l^{(N)})(x)$$

Then, $\forall x \in L_N$,

$$\mathcal{U}_A{}^* W_{k,l}(x) = W_{k,l}(A^{-1}x) = e^{2\pi i(\lambda_k - \lambda_l)} W_{k,l}(x) \tag{2.64}$$

and, in particular,

$$\mathcal{U}_A{}^* W_l(x) = W_l(x)$$

where $W_l = W_{l,l}$, *and*

$$\mathcal{U}_A{}^* W_{k,l}(x) = W_{k,l}(x)$$

whenever $\phi_k^{(N)}$ *and* $\phi_l^{(N)}$ *belong to the same eigenspace of* V_A.

Proof. See [DGI], p.23. Q.E.D.

Notice that we have implicitly established a relation between the degeneracy of the eigenvalues of the quantum propagator V_A and the dimension of the invariant eigenspace of the Koopman operator acting on $L^2(\mathbf{T}^2, d\mu_N)$ (see also [BH], [E1]):

$$\sum_{i=1}^{p(N)} d_i = N, \qquad \sum_{i=1}^{p(N)} d_i^2 = M_N$$

where:

- d_i is the degeneracy order of the i-th eigenvalue of V_A;
- M_N is the number of distinct periodic orbits in $L_N \setminus \{0,0\}$, which coincides with dimE_0;
- E_0 is the invariant eigenspace of $\mathcal{U}_A$ of Proposition 2.1;
- $p(N)$ is the common period of the classical closed orbits in $L_N \setminus \{0,0\}$, which coincides with the quantum period, i.e. $V_A^{p(N)} = \mathrm{Id}$.

As we have seen, $\mathcal{U}_A$ as an operator on $L^2(\mathbf{T}^2, d\mu_N)$, has a point spectrum and each eigenfunction can be written as a sum of functions which are constant on the periodic orbits of L_N (cfr. Proposition 2.6). In the classical limit the non ergodic, zero-entropy, invariant measures μ_N weakly converges to the Lebesgue measure (with positive entropy equal to $\log \lambda$ where λ is the largest eigenvalue of A). This limit measure is ergodic and the Koopman operator has continuous spectrum but for the eigenvalue 1, that is, as we shall see, the Wigner functions weakly converge to constant distributions on the torus.

Now, using Proposition (2.7) and (2.8) one immediately obtain the following result:

Proposition 2.9. *We have the further representation:*

$$\int_{\mathbf{T}^2} f \, d\Omega_N(\phi_k^{(N)}, \phi_l^{(N)}) = \sum_{j=0}^{M_N} \alpha_j(\phi_k^{(N)}, \phi_l^{(N)}) \int_{\mathbf{T}^2} f \, d\mu_{\gamma_j} \tag{2.66}$$

where

$$\mu_{\gamma_0} = \delta_{(0,0)}; \qquad \mu_{\gamma_j} = \frac{1}{p(N)} \sum_{k=0}^{p(N)-1} \delta_{A^k(x)}, \quad x \in \gamma_j, \quad j = 1, \ldots, M_N \tag{2.67}$$

and

$$\alpha_j(k,l) = \alpha_j(\phi_k^{(N)}, \phi_l^{(N)}) = \frac{p(\gamma_j)}{N^2} W_{k,l}|_{\gamma_j} \qquad (2.68)$$

Remark 1. Proposition (2.9) yields an explicit relation between the 'quantum distributions' $d\Omega_N$ and the periodic orbits measures $d\mu_\gamma$ (see the remark on this point in chapter 1). In the corresponding problem for hyperbolic surfaces such a relation is at present at best very unclear [Sa].

Remark 2. The coefficients α_j's may be positive as well as negative, the only condition they have to satisfy being:

$$\sum_{j=0}^{M_N} \alpha_j(k,l) = \begin{cases} 1, & \text{if } l = k; \\ 0, & \text{otherwise.} \end{cases} \qquad (2.69)$$

which follows from the normalisation condition:

$$\int_{T^2} W_{k,l} d\mu_N = \begin{cases} 1, & \text{if } l = k; \\ 0, & \text{otherwise.} \end{cases} \qquad (2.70)$$

This fact makes the present problem quite different from the corresponding problem for ergodic flows on compact manifolds where a positive quantization procedure yields at once a sequence probability measures $d\mu_{\psi_k}$ (see chapter 1) instead of the distributions $d\Omega_N$ found here.

We now state the final result after which Theorem 2.4. easily follows.

Proposition 2.10. *There is a complete orthonormal basis of eigenvectors such that:*

$$\mathcal{H}_N = \phi_0 \bigoplus_{j=0}^{m-1} \bigoplus_{r=0}^{p-1} \phi_{j,r} \qquad (2.71)$$

where $\phi_{j,r}$ with $j = 0,\ldots,m-1$ are the m distinct eigenvectors corresponding to the eigenvalue $e^{-2\pi i r/N}$ of V_A (of constant multiplicity m), whereas $\phi_{j,r}$ with $r = 0,\ldots,p-1$ correspond to the p different eigenvalues, and $V_A \phi_0 = \phi_0$. Moreover, for any pair $(\phi_{k,r}, \phi_{l,r})$ one has the decomposition

$$W_{k,l} = \delta_{k,l} W^0 + W_{k,l}^{\text{mix}} \qquad (2.72)$$

where $W^0 d\mu_N$ is a probability measure which converges in the weak *-topology of $A(T^2)$ to the Lebesgue measure, whereas $W_{k,l}^{\text{mix}}$ vanishes, as $N \to \infty$, $N \in \Gamma$. The speed of convergence is the same as in Theorem 2.4.

Proof. After having constructed a suitable orthonormal basis of eigenvectors and used the explicit action of the operators $\hat{T}(n)$ on such basis, the proof of this result essentially reduces (as in Theorem 2.4) to estimating sums of multiple characters over finite fields (see the Appendix) and we refer the reader to [DGI], Section 4. Q.E.D.

Appendix: Number Theory

In this section p will denote a prime number and $\mathbb{F}_q$ a finite field of characteristic p with $q = p^r$ elements (we will be interested mainly in the case $q = p$, $\mathbb{F}_q = Z_p$).

Let X be a smooth projective absolutely irreducible algebraic curve over $\mathbb{F}_q$ defined by an homogeneus polynomial equation $f(x, y) = 0$ with coefficients in $\mathbb{F}_q$ (whereas the varibles x, y live in the algebraic closure of $\mathbb{F}_q$). Denote moreover by N_k the number of solutions of the above equation in $\mathbb{F}_{q^k}$, so that, in particular, N_1 is the number of points of X in $\mathbb{F}_q$. Observe that one can also characterize the number N_k as the number of fixed points in X of the map F^k where F is the *Frobenius map* defined by

$$F : (x, y) \rightarrow (x^q, y^q) \qquad (A.1)$$

Then, the Weil zeta function is defined by

$$Z(u, X) = \exp \sum_{k=1}^{\infty} \frac{u^k}{k} N_k \qquad (A.2)$$

Moreover, one esily realizes that putting $u = q^{-s}$, it has the Euler product representation

$$Z(s, X) = \sum_h \mathcal{N}(h)^{-s} = \prod_{h \, \text{prime}} (1 - \mathcal{N}(h)^{-s})^{-1} \qquad (A.3)$$

where the sum is over the integral divisors of the algebraic function field $\mathbb{F}_q[x]$ and $\mathcal{N}(h) = q^{deg(h)}$. Futhermore, it is known that $Z(u, X)$ satisfies the functional equation:

$$Z(\frac{1}{u}) = q^{g-1} u^{2-2g} Z(\frac{u}{q}) \qquad (A.4)$$

where g is the genus of the curve X (see, e.g., [Sc]).

Theorem A.1. ([Weil]) $Z(u, X)$ is (a power series expansion of) a rational function of the form

$$Z(u, X) = \frac{P(u)}{(1 - u)(1 - qu)} \qquad (A.5)$$

where $P(u)$ is a polynomial of degree $2g$ and integral coefficients. Moreover,

$$P(u) = \prod_{i=1}^{2g} (1 - \omega_i u) \quad \text{and} \quad |\omega_i| = \sqrt{q}, \, \forall i \qquad (A.6)$$

Notice that (A.6) asserts nothing but the validity of the Riemann hypothesis for curves over finite field, since $Z(q^{-s}) = 0$ only for $Re(s) = 1/2$. Clearly,

$$N_k = \frac{1}{(k-1)!} \frac{d^k}{du^k} (\log Z(u, X))_{u=0} \qquad (A.7)$$

so that, from (A.6),

$$N_k = q^k + 1 - \sum_{i=1}^{2g} \omega_i^k \qquad (A.8)$$

and, in particular, one has the Weil estimate

$$|N_1 - (q+1)| \le 2g\sqrt{q} \qquad (A.9)$$

A (non trivial) consequence is the possibility of estimating certain exponential sums over finite fields.

Example. Consider the case $\mathbb{F}_q = Z_N$ with N prime. Then it may be of some interest to generalize the trivial identity

$$\sum_{x \in Z_N} e^{\frac{2\pi i}{N} x} = 0$$

to the sum

$$\sum_{x \in Z_N} e^{\frac{2\pi i}{N} f(x)}$$

where $f(x)$ is, for instance, a given polynomial. Clearly, the answer is intimately related to the number of solutions of $y - f(x) = 0$ in Z_N.

More generally, we shall consider generalized sums over finite fields of the type:

$$\sum_{x \in \mathbb{F}_q} \chi(f(x))\psi(g(x)) \qquad (A.10)$$

where χ is a non trivial multiplicative character of order $d|q - 1$ of $\mathbb{F}_q$ (i.e. χ^d is equal to the trivial character χ_0), ψ a non trivial additive character of $\mathbb{F}_q$ and $f(x), g(x) \in \mathbb{F}_q[x]$ are given algebraic functions, for instance polynomials, over $\mathbb{F}_q$.

Before we examine some examples, let us give a further definition: let $p \in \mathbb{N}$ and $a \in Z$ be such that $a \ne 0 \pmod{p}$. Then, we say that a is a *quadratic residue* of p if there is $m \in Z$ such that $a = m^2 \pmod{p}$.

Now, given N prime, let once more $\mathbb{F}_q = Z_N$. Then, the only non trivial multiplicative character of order two is the *Legendre symbol* $\chi_2(x) = \left(\frac{x}{N}\right)$ defined as follows (see e.g.[Ap]):

$$\left(\frac{x}{N}\right) = \begin{cases} +1 & \text{if } x \text{ is a quadratic residue;} \\ -1 & \text{otherwise.} \end{cases} \qquad (A.11)$$

Moreover $\left(\frac{0}{N}\right) = 0$ for any $x = 0 \pmod{N}$. The Legendre symbol obviously satisfies the product law

$$\left(\frac{xy}{N}\right) = \left(\frac{x}{N}\right)\left(\frac{y}{N}\right) \qquad (A.12)$$

On the other side, in the present case the group of additive characters is the set

$$\{\psi_a(x) = e^{\frac{2\pi i}{N} ax} ; a \in Z_N\} \qquad (A.13)$$

Theorem A.2. *Let χ, ψ be a multiplicative character $\ne \chi_0$ of order d with $d|(q-1)$, and a non trivial additive character, respectively, of $\mathbb{F}_q$. Let $f(x) \in \mathbb{F}_q[x]$ admit m distinct roots, and let $g(x) \in \mathbb{F}_q[x]$ have degree n. Suppose that either $(d, \deg(f)) = (n, q) = 1$, or, more generally, that the polynomials $y^d - f(x)$ and*

$z^q - z - g(x)$ are absolutely irreducible (i.e. irreducible over any finite algebraic extension of $\mathbb{F}_q$). Then

$$\left| \sum_{x \in \mathbb{F}_q} \chi(f(x))\psi(g(x)) \right| \leq (m + n - 1)q^{1/2} \tag{A.14}$$

Proof. See, e.g., [Sc], page 45.

Remark the above result has been obtained by A. Weil as a consequence of Theorem A.1 and important extensions to cases where f, g are rational functions has been achieved by P. Deligne [De1],[De2].

Let us briefly discuss some consequences of this result.

Example. $f(x) = g(x) = x$, $\chi = \chi_2$ and $\psi = \psi_a$. Then we have the (generalized) *Gauss sum* $G(\psi, \chi)$. If, moreover, $\mathbb{F}_q = Z_N$ we obtain the standard quadratic one

$$g(\psi_a, \chi_2) = \sum_{x \in Z_N} e^{\frac{2\pi i a}{N} x^2} \tag{A.15}$$

and, by direct application of Theorem A.2, $|g(\psi_a, \chi_2)| \leq \sqrt{N}$. More precise information is contained in the following

Lemma A.1.

$$g(\psi_a, \chi_2) = \epsilon_N \sqrt{N} \left(\frac{a}{N} \right) \tag{A.16}$$

where

$$\epsilon_N = \begin{cases} 1, & \text{if } N = 1 (\bmod 4) \\ i, & \text{if } N = 3 (\bmod 4) \end{cases}$$

and

$$\sum_{k=0}^{N-1} \exp \frac{2\pi i}{N}(ak^2 + bk) = \epsilon_N N^{1/2} \left(\frac{a}{N} \right) \cdot \exp\left[-\frac{2\pi i}{N} b^2 (4a)^{-1} \right]$$

if $a \neq 0 (\bmod N)$,

$$\sum_{k=0}^{N-1} \exp \frac{2\pi i}{N}(ak^2 + bk) = N \cdot \delta_b^0$$

if $a = 0 (\bmod N)$.

Example. Consider once more the particular case $\mathbb{F}_q = Z_N$. Setting $f(x) = x^2 - 4ab$, $g(x) = x$, $\chi = \chi_2$ and $\psi = \psi_1$ we then find the *Kloosterman sum*

$$Kl(N, a, b) = \sum_{x \in Z_N} \left(\frac{x^2 - 4ab}{N} \right) e^{\frac{2\pi i}{N} x} = \sum_{x \in Z_N^*} e^{\frac{2\pi i}{N}(ax + bx^{-1})} \tag{A.17}$$

Again, from Theorem A.1, one has the estimate

$$|Kl(N, a, b)| \leq 2\sqrt{N} \tag{A.18}$$

The next two result are useful in reducing to the previous case some sums over cyclic subgroups.

Lemma A.2. *Suppose $d|(q-1)$, then*

$$\sum_{\chi \text{ of order } d} \chi(x) = \begin{cases} d, & \text{if } x \in (\mathbb{F}_q^*)^d \\ 0, & \text{if } x \notin (\mathbb{F}_q^*)^d \ x \neq 0 \\ 1, & \text{if } x = 0 \end{cases}$$

Proof. See [Sc], page 85.

As an immediate consequence, we have:

Lemma A.3. $\forall \lambda \in Z_N^*$, *denote* $\Lambda_\lambda =<\lambda>$ *the cyclic subgroup generated by* λ. *Let* $\#\Lambda_\lambda = \frac{N-1}{m} = p$. *If* $f : Z_N \times Z_N \longrightarrow C$ *is any complex valued function, then*

$$\sum_{s,t=0}^{p-1} f(\lambda^s, \lambda^t) = \frac{1}{m^2} \sum_{j,l=0}^{m-1} \sum_{x,y \in Z_N^*} \chi_j(y)\chi_l(x)f(y,x)$$

where $\{\chi_0, \cdots, \chi_{m-1}\}$ *is a set of multiplicative characters of order* m.

Proof. Clearly $x^m \in \Lambda_\lambda$, $\forall x \in Z_N^*$, because Λ_λ is exactly the set of roots of the polynomial $x^p - 1 = 0$. Moreover, the map $x \longrightarrow x^m \in \Lambda_\lambda$ has multiplicity m. That is,

$$\sum_{s,t=0}^{p-1} f(\lambda^s, \lambda^t) = \frac{1}{m^2} \sum_{x,y \in Z_N^*} f(y^m, x^m)$$

and the result follows immediately from Lemma A.2 because

$$\sum_{x,y \in Z_N^*} f(y^n, x^m) = \sum_{j,l=0}^{m-1} \sum_{x,y \in Z_N^*} \chi_j(y)\chi_l(x)f(x,y)$$

Q.E.D. Using the same technique it is also possible to estimate Kloosterman sums over any cyclic subgroup of Z_N, namely

Proposition A.1 $\forall \lambda \in Z_N^*$, denote $\Lambda_\lambda =<\lambda>$ the cyclic subgroup generated by λ. Let $\#\Lambda_\lambda = \dfrac{N-1}{m} = p$. Then, $\forall a, b \in Z_N^*$

$$|\sum_{x \in \Lambda_\lambda} \exp[\frac{2\pi i}{N}(ax + bx^{-1})]| = |\sum_{s=0}^{p-1} \exp[\frac{2\pi i}{N}(a\lambda^s + b\lambda^{-s})]| \leq C(m)\sqrt{N} \quad (A.18)$$

for some constant $C(m)$ bounded in m.

Proof.

$$\sum_{s=0}^{p-1} \exp\frac{2\pi i}{N}(a\lambda^s + b\lambda^{-s}) =$$

$$\frac{1}{m} \sum_{x \in Z_N} \exp\frac{2\pi i}{N}(ax^m + bx^{-m}) =$$

$$\frac{1}{m} \sum_{j=0}^{m-1} \sum_{x \in Z_N} \chi_j(x) \exp\frac{2\pi i}{N}(ax + bx^{-1})$$

where χ_j $j = 0, \cdots, m-1$ are the multiplicative characters of order m. The assertion now follows from the direct extension of Theorem A.2 to the rational functions of type $ax + bx^{-1}$ (see [De2], pag. 190 and [Sc], pag. 85). Q.E.D.

References.

[AA] V.I.Arnold, A.Avez, *Ergodic Problems in Classical Mechanics*, W.A.Benjamin, New York, 1968.

[Ap] T.Apostol, *Introduction to Analytic Number Theory*, Springer-Verlag, New York 1976.

[AY] V.M.Alexeev, M.V.Yakobson, Symbolic dynamics and hyperbolic dynamical systems, *Phys. Rep* **75**, 287-325 (1981).

[HB] J.H.Hannay, M.V.Berry, Quantization of linear maps on a torus - Fresnel diffraction by a periodic grating, *Physica D* **1**, 267-291 (1980).

[B1] R.Bowen, Periodic orbits for hyperbolic flows, *Amer. J. Math.* **94**, 1-30 (1972).

[B2] R.Bowen, The equidistribution of closed geodesics, *Amer. J. Math.* **94**, 413-423 (1972).

[BalVor] N.L.Balazs, A.Voros, Chaos on the pseudosphere, *Phys. Rep.* **143**, N.3, 109-240 (1986).

[BV] F.Bartuccelli, F.Vivaldi, Ideal Orbits of Toral Automorphisms, *Physica D* **39**, 194 (1989).

[CdV] Y.Colin de Verdiere, Ergodicité et fonctions propres du Laplacien, *Commun. Math. Phys.* **102**, 497-502 (1985).

[DE] M.Degli Esposti, Quantization of the orientation preserving automorphisms of the torus, *Ann.Inst.H.Poincaré* **58**, 323-341 (1993).

[DGI] M.Degli Esposti, S. Graffi, S. Isola, Classical limit of the quantized hyperbolic toral automorphism, *Commun. Math. Phys.*, to appear (1994).

[De1] P.Deligne, La conjecture de Weil I, *Publ. Math. I.H.E.S.* **48**, 273-308 (1974).

[De2] P.Deligne, *Cohomologie Etale*, Lecture Notes in Mathematics **569**, 1977.

[E1] B.Eckhardt, Exact eigenfunctions for a quantized map, *J. Phys.* **A 19**, 1823-1833 (1986).

[E2] B.Eckhardt, Quantum mechanics of classically non-integrable systems, *Phys. Rep.* **163**, 205-297 (1988).

[GVZ] M.Giannoni, A. Voros and J.Zinn-Justin eds., *Chaos and Quantum Physics*, Les Houches 1991, Elsevier Publ. 1992.

[H] H.Hasse, *Number Theory*, Springer-Verlag, Berlin-Göttingen-Heidelberg, 1980.

[He1] D.Hejhal, *Duke Math. J.* **43**, 441-482 (1976).

[He2] D.Hejhal, The Selberg Trace Formula, Vol. I, S.L.N. **548**, (1976).

[He3] D.Hejhal, The Selberg Trace Formula, Vol. II, S.L.N. **1001**, (1980).

[HMR] B.Helffer, A.Martinez and D.Robert, Ergodicité et limite semiclassique, *Commun. Math. Phys.* **131**, 493-520 (1985).

[HPS] M.Hirsch, C.Pugh, M.Schub *Invariant manifolds*, Lect. Notes in Math. **583**, Springer Verlag, Berlin, 1977.

[Hu] H.Huber, Zur analytischen Theorie hyperbolischer Raumforme und Bewegungsgruppen, *Math. Ann.* **138**, 1-26 (1959).

[I] S.Isola, ζ-function and distribution of periodic orbits of toral automorphisms, *Europhysics Letters* **11**, 517-522 (1990).

[K] A.Katok, Entropy and closed geodesics, *Ergodic Theory and Dyn. Syst.* **2**, 339-367 (1982).

[Ka1] N.Katz, *Gauss Sums, Kloosterman Sums, and Monodromy Groups*, Princeton University Press, Princeton, 1988.

[Ka2] N.Katz, Sommes Exponentielles,*Asterisque* **79**, (1980).

[Ke1] J.Keating, Ph.D. thesis University of Bristol, 1989.

[Ke2] J.Keating, Asymptotic properties of the periodic orbits of the cat maps, *Nonlinearity* **4**, 277-307 (1991).

[Ke3] J.Keating, The cat maps: quantum mechanics and classical motion, *Nonlinearity* **4**, 309-341 (1991).

[Ko] J.Korevaar, On Newman's quick way to the prime number theorem, *Mathematical Intelligencer* **4**, 108-115 (1982).

[La] L.D.Landau, E.M.Lifshitz, *Quantum Mechanics*, (3rd edition) Pergamon Press, Oxford, 1965.

[LL] V.F.Lazutkin, *KAM theory and Semiclassical Approximation to Eigenfunctions*, Springer, New York, 1993.

[M] R.Mañe, *Ergodic Theory and Differentiable Dynamics*, Springer, New York, 1987.

[Mar] E.A.Margulis, On some application of ergodic theory to the study of manifolds of negative curvature, *Func. Anal. Appl.* **3**(4), 89-90 (1969).

[PV] I.Percival and F.Vivaldi, Arithmetical properties of strongly chaotic motions, *Physica D* **25**, 105-130 (1987).

[Pa] W.Parry, Equilibrium states and weighted uniform distribution of closed orbits, in Lect. Notes in Math. **1342**, J.C.Alexander Ed. 1988, pp.617-625.

[Po] M.Pollicott, Closed geodesics and zeta functions, in *Ergodic Theory, Symbolic Dynamics and Hyperbolic Spaces*, T.Bedford, M.Keane, C.Series eds., Oxford Science Publ. 1991, pp.153-173.

[PP] W.Parry, M.Pollicott, Zeta functions and the periodic orbit structure of hyperbolic dynamics, *Astérisque* **187-188**, (1990).

[PP1] W.Parry, M.Pollicott, An analogue of the prime number theorem for closed orbits of Axiom A flows, *Annals of Math.* **118**, 573-591 (1983).

[RM] M.Ram Murthy, Artin's Conjecture for Primitive Roots, *Mathematical Intelligencer* **10**, 59-70 (1988).

[Ru1] D.Ruelle, Generalized zeta-functions for expanding maps and Anosov flows, *Inventiones Math.* **34**, 231-242 (1976).

[Ru2] D.Ruelle, *Thermodynamic Formalism*, Enciclopedia of Math. and its Appl., vol. **5**, Addison-Wesley, Reading, Mass. 1978.

[RuSa] Z.Rudnick, P.Sarnak, The behaviour of eigenstates of arithmetic hyperbolic surfaces, preprint, 1993.

[Sa] P.Sarnak, Arithmetic Quantum Chaos, Tel Aviv Lectures 1993 (to appear)

[Schn] A.Schnirelman, Ergodic properties of the eigenfunctions, *Usp. Math. Nauk* **29**, 181-182 (1974).

[Schm] W.Schmidt, *Equations over finite fields. An elementary approach.* Lecture Notes in Mathematics **536**, 1976.

[SI1] Ya.Sinai, The asymptotic behaviour of the number of closed orbits on a compact manifold of negative curvature, *Trans. Amer. Math. Soc.* **73**, 227-250 (1968).

[SI2] Ya.Sinai, Mathematical Problems in the Theory of Quantum Chaos, Tel Aviv Lectures 1990, appeared in *CHAOS/XAOC: Soviet-America Perspectives on Nonlinear Science*, D.K.Campbell ed., New York: American Institute of Physics 1990, p. 395.

[Sm] S.Smale, Differentiable dynamical systems, *Bull. Amer. Math. Soc.* **73**, 747-817 (1967).

[Ta] M.Taylor, *Pseudo-differential operators* Princeton, NJ, Princeton Univ. Press, 1981.

[VN] J.von Neumann, Beweis des Ergodensatzes und des H-Theorems in der Neuen Mechanik, *Zschr.f.Physik* **57**, 30-70 (1929).

[UZ] A.Uribe, S.Zelditch, Spectral statistic on Zoll surfaces, preprint, 1993.

[Wa] P.Walters, *An Introduction to Ergodic Theory*, Graduate Texts in Mathematics, Springer-Verlag, New York, 1982.

[Z] S.Zelditch, Uniform distribution of Eigenfunctions on Compact hyperbolic Surfaces, *Duke Math. J.* **55**, 919-941 (1987).

[Z1] S.Zelditch, On the rate of quantum ergodicity, part 1 and 2, preprints, 1993.

[Z2] S.Zelditch, *Memoirs of A.M.S.* **96**, No. 495, 1992.

ACTION MINIMIZING ORBITS IN HAMILTOMIAN SYSTEMS

John N. Mather
and
Giovanni Forni

CONTENTS

§1. Introduction

The study of the dynamics of area preserving mappings dates from the pioneering work of Poincaré [Po]. Poincaré showed that there is an intimate connection between the dynamics of area preserving mappings and the dynamics of Hamiltonian systems in 2 degrees of freedom.

Consider a Hamiltonian system in 2 degrees of freedom. Such a system is defined by a C^1 function H on a 4 dimensional manifold M provided with a symplectic structure ω, i.e. a closed 2-form whose square vanishes nowhere. Let $X = X_H$ denote the symplectic gradient of H, i.e. the vectorfield on M defined by the condition $dH = i_X\omega$. Let $\{\Phi_t\}_{t\in\mathbf{R}} = \{\Phi_{H,t}\}_{t\in\mathbf{R}}$ denote the flow generated by X. An energy hypersurface $\{H = \text{const.}\}$ is invariant under Φ_t (conservation of energy). If c is a regular value of H, then $\{H = c\}$ is a 3-dimensional submanifold of M. Consider a 2-dimensional surface $S \subset \{H = c\}$ transverse to X, and a trajectory $\{\Phi_t(P)\}_{0\leq t\leq\tau}$ of the flow Φ_H wich begins at $P \in S$ and ends at $\Phi_\tau(P) \in S$, where $\tau > 0$. For Q in a sufficiently small neighborhood $\mathcal{N}$ of P in S, we may find a positive number $\tau(Q)$, depending continuously on Q, with $\tau(P) = \tau$, such that $\Phi_{\tau(Q)}(Q) \in S$. Let $\mathcal{P}(Q) = \Phi_{\tau(Q)}(Q)$.

The mapping $\mathcal{P} : \mathcal{N} \to S$ is called a *Poincaré return mapping* associated to the Hamiltonian system (M, ω, H). It is area preserving, in the following sense. Because $S \subset \{H = c\}$ is transverse to X_H, the restriction of the 2-form ω to S is non-degenerate, i.e. it defines an area form on the 2-dimensional surface S.

The Poincaré return mapping preserves this area form. Moreover, the regularity properties of $\mathcal{P}$ reflect those of H, e.g. if H is C^r ($r \geq 1$), then so is $\mathcal{P}$ (in the region where it is defined).

Poincaré observed that in many cases the dynamical properties of $\mathcal{P}$ are closely related to those of $\{\Phi_t\}_{t \in \mathbf{R}}$. For example, consider the case when P is a fixed point of $\mathcal{P}$. In this case, $\{\Phi_t(P)\}_{0 \leq t \leq \tau}$ is a periodic orbit of the flow. An important question is to know whether a fixed point or a periodic orbit is *stable* in the sense of Lyapunoff. For the fixed point P, this means that there are arbitrarily small neighborhoods of P in S which are invariant under $\mathcal{P}$. Likewise, for the periodic orbit $\{\Phi_t(P)\}_{0 \leq t \leq \tau}$, we could take as definition of Lyapunoff stability that there are aritrarily small neighborhoods of $\{\Phi_t(P)\}_{0 \leq t \leq \tau}$ in the energy hypersurface which are invariant under the flow Φ_t. Obviously, P is Lyapunoff stable for $\mathcal{P}$ if and only if the periodic orbit $\{\Phi_t(P)\}_{0 \leq t \leq \tau}$ is Lyapunoff stable in its energy surface. The famous K.A.M. (Kolmogorov, Arnol'd, Moser) theory gives sufficient conditions for a fixed point P to be Lyapunoff stable for the Poincaré return mapping or, equivalently, for the corresponding periodic orbit to be Lyapunoff stable in its energy hypersurface (cf. Corollary 16.3).

In a similar fashion, Poincaré [Po] and Birkhoff [Bi1] studied the restricted three body problem in classical mechanics, by mean of a Poincaré return map on a Poincaré surface of section, an area preserving map of an annulus. Moser [Mo1] proved stability in the restricted three body problem, a result which follows from the existence of certain invariant closed curves in a suitable Poincaré return map.

The stability of the restricted three body problem is just one of many questions concerning stability vs. randomness in Hamiltonian systems in 2 degrees of freedom in which K.A.M. invariant closed curves of an associated area preserving mapping plays a fundamental role. The book of Moser [Mo1] contains many other examples. Because of the fundamental importance in the dynamics of Hamiltonian systems of invariant curves (and invariant tori in Hamiltonian systems of more than 2 degrees of freedom), there is a large literature on questions of when they exist. See, e.g. Salamon and Zehnder [S-Z] and Herman [He], giving improved estimates.

On 1981, the first author discovered a generalization of certain K.A.M. invariant curves [Ma1] . Similar ideas were discovered by Aubry and his coworkers, independently, around the same time [Au] and [Au-LeD]. The first author's original approach was based on a variational principle introduced by Percival in [Pe1] and [Pe2]. The purpose of these lectures is to describe this generalization of the K.A.M. invariant curves, the minimization procedure used to prove its existence, and related constructions.

The sets which the first author and Aubry constructed can be characterized as supports of invariant measures which minimize the average action in the class of invariant measures with a given rotation number (§20); thus, we will call these sets *action minimizing sets*. The existence of these sets can be proved for a special class of area preserving diffeomorphisms of the annulus (or infinite cylinder) called twist diffeomorphisms. We define this class of diffeomorphisms and give examples in §§2-3.

Then in §6, we prove the existence of action minimizing sets following the original method from [Ma1]. This proof is based on a variational principle introduced by Percival [Pe1], [Pe2]. These sets are more plentiful than K.A.M. invariant curves, because they exist for every twist diffeomorphism and for every rotation number. On the other hand, the existence of K.A.M. invariant closed curves requires further hypotheses. The most complete treatment of the existence theory of K.A.M. invariant closed curves is in Herman's book [He]; we content ourselves in §16 with an overview of some of the known results.

Invariant closed curves in area preserving mappings of an annulus can be classified by their topological placement, i.e. by whether they bound a disk or separate the bottom from the top of an annulus. In analogy with lunar theory, the former curves are called *librational* and the latter are called *rotational*. The theory developed in these notes deals with a generalization of the rotational invariant curves, in the sense that every rotational invariant curve of an area preserving twist diffeomorphism is an action minimizing set, as we will show in §17.

An important result due to G.D.Birkhoff [Bi1], [Bi3] is that every rotational invariant curve of an area preserving twist diffeomorphism f of $\mathbf{S}^1 \times \mathbf{R}$ into itself is the graph of a Lipschitz function $u : \mathbf{S}^1 \to \mathbf{R}$, i.e. it has the form $\{(\theta, u(\theta)) \mid \theta \in \mathbf{S}^1\}$. We discuss this result in §15 and various applications in §17.

In the case of area preserving twist diffeomorphism f of $\mathbf{S}^1 \times \mathbf{R}$ the action minimizing sets mimic invariant curves in the following way: Let $\omega \in \mathbf{R}$ and let $M_\omega^\star$ be the action minimizing set for the rotation number ω. Then there is a closed subset $A_\omega^\star \subset \mathbf{S}^1$ and a Lipschitz mapping $u : A_\omega^\star \to \mathbf{R}$ such that $\Sigma_\omega^\star = \{(\theta, u(\theta)) \mid \theta \in A_\omega^\star\}$. Moreover, $f : \Sigma_\omega^\star \to \Sigma_\omega^\star$ preserves the cyclic order on $\Sigma_\omega^\star$ induced from that on $A_\omega^\star \subset \mathbf{S}^1$. We discuss this result in §14.

Another approach which led to many of the results described in these lectures is due to Aubry and his coworkers. Among the many papers written by Aubry and his coworkers, the one which is most directly relevant to what we dicuss here is the paper of Aubry and Le Daeron [Au-LeD]. Like many of the other papers of Aubry et al. this paper is concerned with the construction of one dimensional crystals (Frenkel-Kontorova model). The mathematical problems which Aubry and Le Daeron study are, however, mathematically equivalent to some of the basic questions on area preserving twist mappings, and their theory provides a useful insight into some of these questions. Their theory leads to the introduction of a class of orbits of area preserving twist mappings, which we call action minimizing orbits. We will discuss the Aubry-Le Daeron theory of action minimizing orbits in §§9-13. In this introduction, we will describe some of the basic results of this theory.

The union Σ of the collection of action minimizing orbits is a closed set of the annulus or cylinder. Associated to each action minimizing orbit, there is a rotation number, which measures, in a certain sense, the rate of advance, with respect to the $\mathbf{S}^1$ factor of a point in the orbit under iteration. Let Σ_ω denote the subset of Σ consisting of action minimizing orbits of rotation number ω. We show in §14 that $\Sigma_\omega^\star \subset \Sigma_\omega$, where $\Sigma_\omega^\star$ is the action minimizing set for the rotation number ω. For irrational ω and a generic area preserving twist diffeomorphism, we have $\Sigma_\omega^\star = \Sigma_\omega$, although in exceptional cases, this equation does not hold. For rational ω and a generic area preserving twist diffeomorphism, $\Sigma_\omega^\star$ consists of a single periodic orbit, but according to the Aubry-Le Daeron theory, Σ_ω contains other orbits as well.

These orbits are heteroclinic to Σ_ω, in the sense of Poincaré, and play an important role in the theory of area-preserving twist diffeomorphisms.

An important application of the theory developed in these lectures concerns destruction of invariant curves. The K.A.M. theorem asserts that, under certain condition, invariant curves persist under perturbations of area preserving twist diffeomorphisms. Since there are several hypotheses for the K.A.M. theorem there are several possible converses. In §18, we discuss a converse of the K.A.M. theorem. Our converse, which generalizes a converse due to Herman [He] shows that the hypotheses concerning the rotation number cannot be relaxed, at least for C^∞ perturbations. We also give a partial converse to Rüssmann's K.A.M. theorem [Rs] in the analytic case, which deals with the optimal condition on the rotation number, when perturbations are required to be small in the analytic topology.

Another application of the theory developed in these lectures concerns orbits in a Birkhoff region of instability of an area preserving twist mapping. In §19, we sketch how to construct certain random or chaotic orbits in such region and various invariant measures (including positive entropy invariant measures) which arise as soon as an action minimizing set of irrational rotation number is not a rotational invariant curve. Complete proofs of the results sketched here are given in [Ma11] and [F].

Finally in §20, we discuss a generalization [Ma12-13], [Mñ] of the theory of action minimizing measures to the case of Lagrangian systems in an arbitrary number of degrees of freedom.

Many people have contributed to the theory of the dynamics of area preserving diffeomorphisms, and more generally to the theory of Hamiltonian systems. In these lectures, we have made no attempt to survey the whole theory. Instead, our intention is to provide an introduction to the theory developed by the first author and S. Aubry, known as "Aubry-Mather theory" . References to work of others is mostly limited to related results and contributions to this theory. We hope these references are sufficiently complete to give a true picture of what is relevant in the theory, but we have not attempted to survey what is known on the dynamics of area preserving mappings.

In §2-14, we give an account, with detailed proofs, of basic results of "Aubry-Mather theory", following the method of Aubry and Le Daeron [Au-LeD], as generalized by Bangert [Ba]. We also compare this with the method of the first author [Ma1]. The rest of these notes were written by the second author, based partly on the first author's lectures at C.I.M.E., and partly on the second author's thesis. Some results are proved, but in order to survey the literature, many results are given without detailed proofs.

The authors wish to thank C.I.M.E. for its hospitality.

§2. Area Preserving Twist Mappings. Definitions.

We let $\mathbf{S}^1$ denote the circle $\mathbf{R}/\mathbf{Z}$. A large part of these lectures is a discussion of the dynamics of a class of mappings of the infinite cylinder $\mathbf{S}^1 \times \mathbf{R}$ onto itself. More generally, we will want to sometimes consider mappings of an open subset U of $\mathbf{S}^1 \times \mathbf{R}$ into $\mathbf{S}^1 \times \mathbf{R}$.

For convenience, we will use the following notation throughout: we will let θ (mod.1) denote the standard parameter of $\mathbf{S}^1$ and x the corresponding parameter of its universal cover, i.e. $\mathbf{R}$. We will let y denote the standard parameter of the second factor of $\mathbf{S}^1 \times \mathbf{R}$ so that (θ, y) is a global system of coordinates for $\mathbf{S}^1 \times \mathbf{R}$ and (x, y) is a global system of coordinates of its universal cover.

Now we describe the class of mappings which we will consider. We let J denote the set of pairs (U, f), whih the following properties. We require that U be an open subset of $\mathbf{S}^1 \times \mathbf{R}$ which intersects each vertical line $\theta \times \mathbf{R}$ in an open interval, which may be infinite at one or both ends. We require that f be a homeomorphism of U onto an open subset fU which also intersects each vertical line in an open interval. We require that f be orientation preserving and area preserving.

We require furthermore that f be exact. To explain this condition, we consider a Jordan curve Γ in U which is homotopically non-trivial in U and has zero area. We let $C = \mathbf{S}^1 \times y$, where $y << 0$, so that both Γ and $f\Gamma$ lie above C. The requirement that f be exact is that the area above C and below Γ is the same as the area above C and below $f\Gamma$. Since f is area preserving and orientation preserving this condition is independent of the choice of Γ.

Finally, we impose a monotone twist condition. Let $\pi_1 : \mathbf{S}^1 \times \mathbf{R} \to \mathbf{S}^1$ denote the projection on the first factor. A monotone twist condition is that, for each $\theta \in \mathbf{S}^1$, the mapping

$$\pi_1 f : U \cap (\theta \times \mathbf{R}) \to \mathbf{S}^1$$

is a local homeomorphism. Since the domain is an open interval and the range is $\mathbf{S}^1$, this condition amounts to saying that $\pi_1 f | U \cap (\theta \times \mathbf{R})$ is monotone increasing or monotone decreasing. If $\pi_1 f | U \cap (\theta \times \mathbf{R})$ is monotone incresing for all $\theta \in \mathbf{S}^1$, we will say that f satisfies a *positive* monotone twist condition. In the other case, we will say that f satisfies a *negative* monotone twist condition. Either of these cases can be reduced to the other, obviously. We impose the positive monotone twist condition.

Thus, J is the set of pairs (U, f), where U and fU are as above and f is exact area preserving, orientation preserving homeomorphism which satisfies a positive monotone twist condition, and is homotopic to the inclusion mapping.

We will also consider, for each positive integer r and for $r = \infty$, the subset J^r consisting of all pairs $(U, f) \in J$ such that f is a C^r diffeomorphism of U onto fU and $\pi_1 f | \theta \times \mathbf{R}$ has everywhere positive derivative for each $\theta \in \mathbf{S}^1$.

In the next two sections we give some examples of elements of J.

§3. Area Preserving Twist Mappings. Examples.

In an elementary expository article [Bi2], G.D. Birkhoff discussed billiards in a convex region as a simple and appealing example of a Hamiltonian system. For our discussion we will consider a convex, bounded region R whose boundary ∂R is C^2. Billiards in R means the dynamics of a particle which moves at constant velocity in R and is reflected by the boundary, according to the rule angle of incidence equals angle of reflection.

There is a simple way to prove the existence of periodic orbits in R, discussed by Birkhoff in [Bi2]. Among all convex q-gons inscribed in R, consider one of maximum perimeter. This exists by a simple compactness argument. An elementary and well known variational argument shows that two sides incident to the same vertex make the same angle with respect to the boundary, for such a maximal q-gon. Thus, a particle which travels at constant speed along the boundary of a maximal q-gon describes an orbit of billiards. Such a trajectory is obviously periodic.

More generally, Birkhoff calls a polygon inscribed in R *harmonic* if at each vertex, the two sides incident to that vertex make the same angle with respect to the boundary. Obviously, harmonic polygons correspond to periodic orbits.

As Birkhoff remarked, the dynamics of billiards in R can largely be reduced to the dynamics of an associated area preserving mapping. If $P \in \partial R$ and $0 < \phi < \pi$, consider the ray $r = r(P, \phi)$ beginning at P and making an angle ϕ with ∂R, counted in the counterclockwise direction starting from ∂R.

Extend r until it meets ∂R again. Let P' be the point other than P where r meets ∂R and let ϕ' be the angle which r makes at P' with ∂R, counted in the clockwise direction starting from ∂R.

Let f be the mapping which associates (P', ϕ') to (P, ϕ). Thus, f is a mapping of $\partial R \times (0, \pi)$ into itself. The dynamics of f is closely related to the dynamics of billiards in R. For example, periodic orbits of f correspond to periodic trajectories of the billiards.

Let s denote the arc length parameter of ∂R, with the counterclockwise direction taken as positive. Let L denote the total arc length of ∂R. Let $h(s, s') = -\|P_s - P_{s'}\|$, for $s \leq s' \leq s + L$, where P_s denotes the point of ∂R with parameter value s and $\|\cdot\|$ denotes the Euclidean norm on the plane. (The sign is chosen so as to agree with sign convention which we use throughout these lectures.) It is easy to see that the relation

$$(3.1) \qquad\qquad f(P_s, \phi) = (P_{s'}, \phi')$$

is equivalent to the two equations

$$(3.2) \qquad\qquad \begin{cases} y = -\partial_1 h(s, s') \\ y' = \partial_2 h(s, s') \end{cases}$$

if $y = -\cos \phi$ and $y' = -\cos \phi'$. In other words, given four numbers s, s', ϕ, ϕ' satisfying $s \leq s' \leq s + L$ and $0 \leq \phi \leq \pi$, $0 \leq \phi' \leq \pi$, these numbers satisfy (3.1) if and only if they satisfy (3.2). Here, ∂_1 and ∂_2 denote the first partial derivatives with respect to the first and second variable.

From this it is easy to see that f preserves the 2-form $ds \wedge dy = \sin\phi\, ds \wedge d\phi$. Indeed, by (3.2), $ds \wedge dy = -\partial_{12}\, h(s, s')\, ds \wedge ds' = ds' \wedge dy'$. Thus, f is orientation preserving and area preserving (for the measure $ds\, dy = \sin\phi\, ds\, d\phi$). Exactness is obvious in this case. The positive monotone twist condition is also obvious.

A second important example arises from the Frenkel-Kontorova model of solid state physics. Let $h(x, x') = (x - x')^2/2 + u(x)$, where u is a sufficiently smooth real valued function of one real variable.

The Frenkel-Kontorova model concerns the case $u(x) = k \sin 2\pi x$, where k is a real parameter. In the model, a crystal is represented as bi-infinite sequence $x = (..., x_i, ...)$ of real numbers (which are thought of as the positions of the atoms of the crystal), which minimize the energy $\sum_{i=-\infty}^{\infty} h(x_i, x_{i+1})$. Since the sum $\sum_{i=-\infty}^{\infty} h(x_i, x_{i+1})$ is rarely convergent, the energy is not defined. However, the notion of minimizing the energy can be defined as follows: x minimizes the energy if for every bi-infinite sequence $\xi = (..., \xi_i, ...)$ such that $\sum_{-\infty}^{\infty} |\xi_i - x_i| < \infty$, we have

$$\sum_{-\infty}^{\infty} (h(\xi_i, \xi_{i+1}) - h(x_i, x_{i+1})) \geq 0$$

and the sum on the left is absolutely convergent.

To the Frenkel-Kontorova model, it is possible to associate an exact area preserving, orientation preserving twist mapping f defined by $f(x, y) = (x', y')$ if and only if

$$(3.3) \qquad \begin{cases} y = -\partial_1 h(x, x') \\ y' = \partial_2 h(x, x') \ . \end{cases}$$

To be explicit, we may solve these equations and find that f is given by

$$(3.4) \qquad x' = x + y - du(x)/dx \qquad y' = y - du(x)/dx \ .$$

In this model, a bi-infinite sequence $x = (..., x_i, ...)$ is called a *configuration* and such a configuration is said to be *stationary* if

$$\frac{\partial}{\partial x_i}(h(x_{i-1}, x_i) + h(x_i, x_{i+1})) = 0 \ , \quad i \in \mathbf{Z} \ .$$

This condition is obtained by formally differentiating the infinite sum

$$\sum_{-\infty}^{\infty} h(x_i, x_{i+1})$$

and setting the result equal to zero. Of course, this procedure has no mathematical meaning, but it is easily verified that a crystal in the model (also called a *minimal configuration*) is stationary in this sense.

If $x = (..., x_i, ...)$ is a stationary configuration and

$$y_i = -\partial_1 h(x_i, x_{i+1}) = \partial_2 h(x_{i-1}, x_i) \ ,$$

then $f(x_i, y_i) = (x_{i+1}, y_{i+1})$, i.e. $(..., (x_i, y_i), ...)$ is an orbit of the mapping f. This correspondence between orbits of f and minimal configurations of h is one-one.

Since $f(x + 1, y) = f(x, y) + (1, 0)$, we obtain a mapping of $\mathbf{S}^1 \times \mathbf{R}$ into itself which is in the class J^∞, introduced in the previous section. Thus, the theory developed in these lectures applies to the Frenkel-Kontorova model. In fact, Aubry et al. developed their theory in the context of the Frenkel-Kontorova model, and we will use many ideas from their theory in these lectures.

In dynamics, one wishes to understand all orbits. The interest in the minimal orbits (associated to the minimal configurations) arises from the fact that the minimal orbits form a class of orbits which can be fairly thoroughly understood. This is in contrast to the situation for general orbits of even simple area preserving mappings, such as the mapping associated to the Frenkel-Kontorova model. The orbit structure of general area preserving mappings defies complete understanding. On the other hand, in the Frenkel-Kontorova model, the crystals (minimal configurations) are the main object of study, and Aubry and Le Daeron [Au-LeD] gave a fairly complete picture of their structure.

The mapping associated to the Frenkel-Kontorova model was extensively studied numerically (and non-rigorously) by physicists such as Chirikov, Greene, Percival; starting in the seventies. We will not discuss these numerical studies in these lectures however. The reader may consult the book of Lichtenberg and Lieberman [L-L]. We will refer to a mapping of the form (3.4) with $u(x) = k \sin 2\pi x$ as a Chirikov mapping.

§4. Birkhoff Normal Form.

Let f be a C^r mapping of an open set in $\mathbf{R}^2$ in $\mathbf{R}^2$. Let P be a fixed point of f. Let λ and μ be the eigenvalues of $df(P)$. We will suppose that f is area preserving and orientation preserving. Then $\lambda\mu = 1$. Let $\tau = \lambda + \mu$ denote the trace of $df(P)$.

It is traditional to classify fixed points of area preserving mappings according to three cases: if $|\tau| > 2$, the fixed point is said to be *hyperbolic*. If $|\tau| = 2$, the fixed point is said to be *parabolic*. If $|\tau| < 2$, the fixed point is said to be *elliptic*.

If one of the eigenvalues, say λ, is imaginary, then the other is the complex conjugate of it. Then $|\lambda|^2 = \lambda\overline{\lambda} = \lambda\mu = 1$, and $|\mu|^2 = 1$. Clearly, $|\tau| = |\lambda + \mu| < 1$, so in this case the fixed point is elliptic. If, on the other hand, both eigenvalues are real, then $|\tau| = |\lambda + \mu| \geq 2$, so the fixed point is parabolic or hyperbolic.

In the case that the fixed point is hyperbolic, the dynamics of f is fairly completely described by the Hartman-Grossman theorem which asserts that the germ of f at P is topologically conjugate to its linear part, i.e. there exists a homeomorphism h of an open neighborhood U of P in $\mathbf{R}^2$ onto an open neighborhood hU of P such that $h \circ df(P) = f \circ h$, where both sides are defined. See e.g. [P-deM].

In contrast, the case when the fixed point is parabolic is complicated, but it does not occur generically, and we will not discuss it.

A partial understanding of the case when the fixed point is elliptic may be obtained through Birkhoff normal form, which we will discuss in this section.

Birkhoff Normal Form Theorem [Bi1]. *Suppose that P is an elliptic fixed point of a C^∞ area preserving mapping f of an open subset of the plane into the plane. There exists a C^∞ local coordinate system ξ, η, centered at P with respect to which f has the form $f(\zeta) = \zeta'$, where*

$$\zeta' = \zeta \exp 2\pi i(\beta_0 + \beta_1\rho^2 + \ldots + \beta_N\rho^{2N}) + O(\rho^k)$$

and $k = 2N + 2$ or $2N + 3$. Here, $\zeta = \xi + i\eta$ denote the complex coordinate associated to the coordinate system ξ, η, and $\rho = (\xi^2 + \eta^2)^{1/2}$. If an eigenvalue λ of $df(P)$ is not a root of unity, then we may take $N = \infty$ (with reminder $O(r^\infty)$). If λ is a primitive q-root of unity, then we may take $k = q$ and $N = [q - 2/2]$. Furthermore the coordinate system ξ, η may be chosen to preserve the area form, i.e. $d\xi \wedge d\eta = dx \wedge dy$, where x, y are the standard coordinates on $\mathbf{R}^2$.

We will refer to the mapping

$$N(\zeta) = \zeta \exp 2\pi i(\beta_0 + \beta_1\rho^2 + \ldots + \beta_N\rho^{2N})$$

as the *normal form* of f. Thus, f is the sum of the normal form and a reminder term which is no bigger than $O(\rho^k)$, where $k = \infty$ if an eigenvalue λ of $df(P)$ is not a root of unity and $k = q$ if an eigenvalue is a qth root of unity.

Since β_0, β_1, ... are real numbers, $|\exp 2\pi i(\beta_0 + \beta_1\rho^2 + \ldots)| = 1$. Thus, the circles $\rho = $ const. are invariant under the normal form N. Clearly, $\exp 2\pi i\beta_0$ is one of the eigenvalues of $df(P)$. The remaining β_i's are called the *Birkhoff invariants* of

If at least one of the Birkhoff invariants is not zero, then f is a twist mapping in a sufficiently small punctured neighborhood of P. More precisely, let $\theta = \tan^{-1} \eta/\xi$. Then $\Phi = (\theta, \rho^2)$ maps $U \setminus \{P\}$ into $\mathbf{S}^1 \times \mathbf{R}$, where U is a sufficiently small open neighborhood of P in $\mathbf{R}^2$. Let $W = \Phi(U \setminus \{P\}) \subset \mathbf{S}^1 \times \mathbf{R}$. Let $g = \Phi \circ f \circ \Phi^{-1} : W \to \mathbf{S}^1 \times \mathbf{R}$. If one of the Birkhoff invariants is non-vanishing and the first non-vanishing Birkhoff invariant is positive, then $(W, g) \in J^\infty$, provided that U is chosen to be a sufficiently small neighborhood of P. On the other hand, if the first non-vanishing Birkhoff invariant is negative, then $(gW, g^{-1}) \in J^\infty$.

This is our third example of an area preserving twist mapping.

§5. The Variational Principle.

Let $(U, f) \in J$ (see §2). Let $\tilde{U} = \pi^{-1}U$, where $\pi : \mathbf{R}^2 \to \mathbf{S}^1 \times \mathbf{R} = (\mathbf{R}/\mathbf{Z}) \times \mathbf{R}$ denotes the projection. Let $\tilde{f} : \tilde{U} \to \mathbf{R}^2$ be a lift of f to the universal cover. Let $V \subset \mathbf{R}^2$ be the set of (x, x') such that there exists $y, y' \in \mathbf{R}$ with $(x, y) \in \tilde{U}$ and $\tilde{f}(x, y) = (x', y')$. Clearly, V is open in $\mathbf{R}^2$. Note that if $(x, x') \in V$, there is only one pair (y, y') which satisfies this condition, by the twist condition. Moreover, (y, y') depends continuously on (x, x').

From the twist condition, it follows that V intersects each vertical line in $\mathbf{R}^2$ and each horizontal line in $\mathbf{R}^2$ in an open interval. In particular V is simply connected. From the area preserving property of f, it follows that $y' \, dx' - y \, dx$ is a closed 1-form (in the sense of distributions) and therefore there is a C^1 real valued function $h = h(x, x')$ on V such that

$$(5.1) \qquad\qquad dh = y' \, dx' - y \, dx \ .$$

Moreover, if $(U, f) \in J^r$, then h is of class C^{r+1} on V.

We will call h the *variational principle* associated to (U, f). We will extend h to all of $\mathbf{R}^2$ by the convention that $h(x, x') = +\infty$ when $(x, x') \notin V$.

Note that the condition (5.1) is equivalent to the equations

$$(5.2) \qquad\qquad \begin{cases} y = -\partial_1 h(x, x') \\ y' = \partial_2 h(x, x') \ . \end{cases}$$

Thus the variational principle could also be defined by the condition that (5.2) holds if and only if $\tilde{f}(x, y) = (x', y')$. The variational principle associated to (U, f) is unique up to addition of a constant.

We have already seen two examples of this variational principle. For billiards this is expressed by the equivalence of (3.1) and (3.2). We defined the Chirikov mapping by (3.3).

The variational principle satisfies the following periodicity condition: V is invariant under the traslation $(x, x') \to (x + 1, x' + 1)$ of the plane, and

$$(5.3) \qquad\qquad h(x + 1, x' + 1) = h(x, x') \ .$$

This is an easy consequence of exactness, which means that

$$\int_\gamma y' \, dx' - y \, dx \ = \ 0 \ ,$$

where the integral is taken over a curve which goes once around the cylinder.

It satisfies the following monotonicity conditions: in V, $\partial_1 h(x, x')$ is a strictly decreasing function of x' and $\partial_2 h(x, x')$ is a strictly decreasing function of x. The first follows from $y = -\partial_1 h(x, x')$ and the positive monotone twist condition (x' is a strictly increasing function of y). The second follows from $y' = \partial_2 h(x, x')$ and the fact that f^{-1} satisfies a negative monotone twist condition (x is a strictly decreasing function of y').

If $(U, f) \in J^1$, then the stronger condition

(5.4)
$$\partial_{12} h(x, x') < 0 , \quad (x, x') \in V$$

is satisfied.

We will also need the integrated form of (5.4), which is also true for $(U, f) \in J$:

(5.5)
$$h(\xi, \xi') + h(x, x') - h(\xi, x') - h(x, \xi') < 0 ,$$
$$\text{if } \xi < x \text{ and } \xi' < x' .$$

It follows easily from the twist condition that the domain V where h is finite has the following form: there exist functions β_-, β_+ such that $\beta_\pm(t+1) = \beta_\pm(t)+1$, both are increasing (i.e. $s \leq t$ implies $\beta_\pm(s) \leq \beta_\pm(t)$), β_- (resp.β_+) is everywhere finite or identically $-\infty$ (resp.$+\infty$), and V is the region above graph β_- and below graph β_+. Here graph $\beta_\pm$ is empty if $\beta_\pm = \pm\infty$ and graph $\beta_\pm = \{(x, x') \mid \beta_\pm(x-0) \leq x' \leq \beta_\pm(x+0)\}$ if $\beta_\pm$ is finite.

§6. Existence of Action Minimizing Sets.

Let $(U, f) \in J$, let h be the variational principle associated to it and let V be the domain on which h is finite. Let

$$h^*(x, x') = \liminf_{(\xi, \xi') \to (x, x')} h(x, x') .$$

Then $h^* = h$ except on the frontier of V, since h is continuous on V and $+\infty$ elsewhere. Moreover, h^* is lower semi-continuous.

Let Y_1 denote the set of mappings $\phi : \mathbf{R} \to \mathbf{R}$ which are increasing (i.e.$s \leq t$ implies $\phi(s) \leq \phi(t)$) and satisfying the periodicity condition $\phi(t+1) = \phi(t)+1$. Let Y denote Y_1 modulo the following identifications: $\phi \sim \psi$ if there exists $a \in \mathbf{R}$ such that $\phi(t) = \psi(t + a)$ at all but at most countably many t.

For $\omega \in \mathbf{R}$ and $\phi \in Y_1$, let

(6.1) $$F_\omega(\phi) = \int_a^{a+1} h^*(\phi(t), \phi(t + \omega)) \, dt .$$

To show that this integral exists (possibly with the value $+\infty$), we will show that the integrand is bounded below. First, $h^*(x, x') > -\infty$ everywhere. This is obvious except when (x, x') is in the frontier of V. Suppose, for example, that $(\xi, x') \in$ graph β_+. It is easy to see that there exists $x > \xi$ and $\xi' < x'$ such that (ξ, ξ'), (x, ξ') and (x, x') are in V. Then (5.5) guarantees that

$$h^*(\xi, x') \geq h(\xi, \xi') + h(x, x') - h(x, \xi') .$$

The case of a point in graph β_- may be treated similarly. Thus, $h^*(x, x') > -\infty$ everywhere.

Since $\phi(t) + \omega - 1 \leq \phi(t + \omega) \leq \phi(t) + \omega + 1$, and $h^*(x + 1, x' + 1) = h^*(x, x')$, the integrand of (6.1) is bounded below as asserted. Thus, the integral (6.1) exists, although it may be $+\infty$.

The integral (6.1) is independent of the choice of $a \in \mathbf{R}$, since $h^*(x+1, x'+1) = h^*(x, x')$ and $\phi(t + 1) = \phi(t) + 1$. Furthermore, if ϕ and ψ define the same element in Y, then $F_\omega(\phi) = F_\omega(\psi)$. Thus, we have defined a function

$$F_\omega : Y \to \mathbf{R} \cup \{+\infty\} ,$$

which we call *Percival's Lagrangian* (compare [Ma1]).

For $\phi \in Y_1$, let graph $\phi = \{(x, x') \,|\, \phi(x - 0) \leq x' \leq \phi(x + 0)\}$. For $\phi, \psi \in Y_1$, let $d_1(\phi, \psi)$ denote the Hausdorff distance between graph ϕ and graph ψ, i.e.

$$d_1(\phi, \psi) = \max \left\{ \sup_P \inf_Q \|P - Q\| , \ \sup_Q \inf_P \|P - Q\| \right\} ,$$

where P ranges over graph ϕ, Q ranges over graph ψ and $\|\cdot\|$ denotes the Euclidean distance. For $[\phi], [\psi] \in Y$, let

$$d([\phi], [\psi]) = \inf \{ d_1(\phi, \psi) \,|\, \phi \in [\phi] , \ \psi \in [\psi] \} .$$

It is easy to verify that d is a metric on Y and Y is compact with respect to this metric.

The topology associated with this metric is the topology of almost everywhere convergence.

Let λ denote the Lebesgue measure on $\mathbf{R}$. If $\phi \in Y_1$, then $\phi_* \lambda$ is a measure on $\mathbf{R}$ which is invariant under the translation $t \to t + 1$. Let $\pi : \mathbf{R} \to \mathbf{R}/\mathbf{Z}$ denote the projection. Then $\pi_*(\phi_* \lambda|[a, a+1))$ is a probability measure on $\mathbf{R}/\mathbf{Z}$ which we will denote (by abuse of terminology) by $\phi_* \lambda$. This measure is independent of the choice of $a \in \mathbf{R}$ and depends only on the equivalence class of ϕ in Y. In fact, $[\phi] \to \phi_* \lambda$ is a homeomorphism of Y onto the space of probability measures on $\mathbf{R}/\mathbf{Z}$, provided with the vague topology.

It is easy to check that F_ω is lower semi-continuous with respect to the topology of almost everywhere convergence. (Compare [Ma1] where a detailed proof that F_ω is continuous was given under a slightly stronger hypothesis). Hence:

Theorem 6.1. *There exists a minimizer ϕ_ω of F_ω.*

It may happen, however, that $F_\omega \equiv +\infty$.

The connection with the dynamics comes from the Euler-Lagrange equation:

Theorem 6.2. *If $\phi = \phi_\omega$ is a minimizer of F_ω, and $\phi(t + \omega \pm 0) \in V$ for all $t \in \mathbf{R}$, then the Euler-Lagrange equation*

$$\partial_2 h(\phi(t - \omega \pm 0), \phi(t \pm 0)) + \partial_1 h(\phi(t \pm 0), \phi(t + \omega \pm 0)) = 0$$

is satisfied for all $t \in \mathbf{R}$.

Since ϕ may be discontinuous, it is essential to use the limits

$$\phi(t + 0) = \lim_{s \to t^-} \phi(s) \quad \text{and} \quad \phi(t - 0) = \lim_{s \to t^+} \phi(s) .$$

Proof. Consider a 1-parameter family ϕ_s with $\phi_0 = \phi$. We have

$$\frac{d}{ds} F_\omega(\phi_s)|_{s=0} = \int_a^{a+1} \frac{d}{ds} h(\phi_s(t), \phi_s(t + \omega))|_{s=0} \, dt =$$

$$= \int_a^{a+1} \{\partial_1 h(\phi(t), \phi(t + \omega)) + \partial_2 h(\phi(t - \omega), \phi(t))\} \frac{d}{ds} \phi_s(t)|_{s=0} \, dt ,$$

at least formally, where we have used $\phi_s(t + 1) = \phi_s(t) + 1$ and $h(x + 1, x' + 1) = h(x, x')$ to make the change of variables. The problem is to choose the family so that $\phi_s \in Y_1$ and so that the formal operations above are justified.

First, we consider the case when $t - \omega$, t and $t + \omega$ are points of continuity of ϕ, and ϕ is not constant in any interval containing t. In this case, we let u_s be a 1-parameter family of diffeomorphisms of $\mathbf{R}$ with $u_0 = $ identity. We suppose that u_s dependes infinitely differentiably on all variables and $u_s(x + 1) = u_s(x) + 1$. We set $\phi_s = u_s \circ \phi$. We may suppose that $u_s(x) = x$ for $x \notin [\phi(t) - \delta, \phi(t) + \delta] + \mathbf{Z}$. Then the formal argument above applies and $d\phi_s/ds|_{s=0}$ vanishes outside $\phi^{-1}([\phi(t) - \delta, \phi(t) + \delta]) + \mathbf{Z}$. By choosing u_s appropriately we may assume that $d\phi_s/ds|_{s=0}$ always has the same sign in the non-vanishing region. By the continuity assumptions and the assumption that ϕ is not constant in any interval containing t, it will be the case that $\partial_2 h(\phi(t-\omega), \phi(t)) + \partial_1 h(\phi(t), \phi(t+\omega))$ always has the same sign in the region where $d\phi_s/ds|_{s=0}$ does not vanish. Hence, the Euler-Lagrange equation is satisfied in this case.

Next, we drop the assumption that ϕ is not constant in any interval containing t, but we retain the other assumptions. We let $[t_0, t_1]$ be the maximal interval containing t on which ϕ is constant. From the monotonicity properties of ϕ related to (5.4), it follows that

$$(6.2) \qquad \partial_2 h(\phi(t-\omega), \phi(t)) + \partial_1 h(\phi(t), \phi(t+\omega)) = E(t)$$

is a decreasing function in $[t_0, t_1]$ and is constant in this interval only if both $\phi(t-\omega)$ and $\phi(t+\omega)$ are constant in it.

If we try to apply the same argument as before, we may run into the problem that $E(t)$ changes sign in the interval $[t_0, t_1]$. If it doesn't, we may apply the same argument as before. If it does, we choose t^* such that $E(\tau) \geq 0$ for $\tau \leq t^*$ and $E(\tau) \leq 0$ for $\tau \geq t^*$ (and $\tau \in [t_0, t_1]$ in both cases). We choose u_s as above and for definiteness we suppose that $du_s(x)/ds|_{s=0} > 0$ where it does not vanish. In this case, however, we define $\phi_s = u_s \circ \phi$ in $[t^*, t+1/2]$ and $\phi_s = u_{-s} \circ \phi$ in $(t-1/2, t^*)$, and extend it by periodicity.

We still have $\phi_s \in Y_1$, for $s \geq 0$ (but not for $s < 0$; the order preserving property fails). Moreover, $dF_\omega(\phi_s)/ds|_{s=0} < 0$, so we obtain a contradiction to the assumption that ϕ minimizes F_ω. Thus, we again obtain the Euler-Lagrange equation in this case. (Note that this argument fails if ϕ maximizes F_ω).

Thus we have proved the Euler-Lagrange equation when $t - \omega$, t, $t + \omega$ are points of continuity of ϕ. The set of t satisfying these conditions is dense in $\mathbf{R}$. By passing to the limit, we obtain the conclusion of Theorem 6.2. $\qquad\square$

Suppose that the hypotheses of Theorem 6.2 holds. Set

$$\eta^\pm(t) = -\partial_1 h(\phi(t \pm 0), \phi(t + \omega \pm 0)) = \partial_2 h(\phi(t - \omega \pm 0), \phi(t \pm 0)) .$$

Let $M_\omega^* = M_\omega^*(f) = \{(\phi(t \pm 0), \eta^\pm(t)) \,|\, t \in \mathbf{R}\}$. Then $M_\omega^* \subset \tilde{U}$ (the universal covering space of U; cf. §5) and is invariant under $\tilde{f}$. These assertions follow from the definition of V and h (§5). Obviously M_ω^* is invariant under the translation $(x, y) \to (x+1, y)$. We let $\Sigma_\omega^* = \Sigma_\omega^*(f) \subset U$ denote the projection of M_ω^*. Obviously, Σ_ω^* is invariant under f. We call Σ_ω^* (or M_ω^*) an action minimizing set of rotation number ω for f (or $\tilde{f}$).

To finish our discussion of the existence of action minimizing sets, we need to describe conditions under which the hypothesis of Theorem 6.2 that $(\phi_\omega(t \pm 0), \phi_\omega(t + \omega \pm 0)) \in V$ for all $t \in \mathbf{R}$ is satisfied.

In the case of the Chirikov mapping $V = \mathbf{R}^2$, so this condition is satisfied trivially. In this case, we may therefore conclude that an action minimizing set Σ_ω^* exists for every real number ω.

More generally, $V = \mathbf{R}^2$ and consequently Σ_ω^* exists for every real ω when $U = fU = \mathbf{S}^1 \times \mathbf{R}$ and

$$\pi_1 \tilde{f}(x,y) \to \pm\infty , \quad \text{as } y \to \pm\infty ,$$

where $\pi_1 : \mathbf{R}^2 \to \mathbf{R}$ denotes the projection on the first factor. When this condition is satisfied, we will say that f *twists each end of the cylinder infinitely*.

To fit billiards into our scheme, we introduce the coordinate $x = s/L$, where s is the parameterization of the boundary by arc-length and L is the total arc-length of the boundary. Then the mapping associated to billiards is area preserving for the area form $dx \wedge dy$, where $y = -\cos\phi$, as before. In this case $U = \mathbf{S}^1 \times (-1,1)$ and $V = \{(x,x') \in \mathbf{R}^2 \,|\, x < x' < x+1\}$. If $\omega < 0$ or $\omega > 1$, then $F_\omega \equiv +\infty$. Also when $\omega = 0$ or $\omega = 1$, no ϕ satisfies the hypothesis of Theorem 6.2.

When $0 < \omega < 1$, the minimizing ϕ satisfies the hypothesis of Theorem 6.2. Obviously, there exist ϕ such that $F_\omega(\phi) < +\infty$, so $F_\omega(\phi_\omega) < +\infty$, where ϕ_ω is a minimizing ϕ. Consequently, $(\phi_\omega(t \pm 0), \phi_\omega(t + \omega \pm 0)) \in \overline{V}$, for all $t \in \mathbf{R}$. Thus, it suffices to show that $(\phi_\omega(t \pm 0), \phi_\omega(t + \omega \pm 0))$ is never in the boundary of V, i.e. it suffices to exclude the possibility $\phi_\omega(t + \omega \pm 0) = \phi_\omega(t \pm 0)$ and $\phi_\omega(t + \omega \pm 0) = \phi_\omega(t \pm 0) + 1$. For example, suppose that $\phi_\omega(t + \omega + 0) = \phi_\omega(t + 0)$. From the monotonicity and the periodicity of ϕ_ω, it follows that there exists an integer $n \geq 2$ such that $\phi_\omega(t + n\omega + 0) > \phi_\omega(t + 0)$. Let n be the smallest such integer. Let $\overline{x} = \phi_\omega(t + (n-2)\omega + 0) = x = \phi(t + (n-1)\omega + 0)$ and $x' = \phi_\omega(t + n\omega + 0)$. It is easy to check that $\partial_2 h(\overline{x}, x) + \partial_1 h(x, x') < 0$. Using an argument similar to that used in the proof of Theorem 6.2, we may then produce a 1-parameter family ϕ_s, $0 \leq s \leq \epsilon$, $\phi_s \in Y_1$, $\phi_0 = \phi_\omega$, such that $dF_\omega(\phi_s)/ds < 0$. This contradicts the assumption that ϕ_ω minimizes F_ω. The other cases may be treated similarly and we obtain that ϕ_ω satisfies the hypothesis of Theorem 6.2.

We may therefore conclude that in the case of billiards, there exists an action minimizing set M_ω^* for every $0 < \omega < 1$.

We did not feel that it was necessary to give the complete details in the argument above, since they were given in a more general setting in [Ma1]. There we showed that if $(U, f) \in J$ and $fU = U$, then M_ω^* exists for every ω in a certain range $\omega_- < \omega < \omega_+$, where ω_- and ω_+ are the rotation numbers of the mappings β_- and β_+ defining the boundary of V. (See the end of §5). In the case of billiards, $\omega_- = 0$ and $\omega_+ = 1$.

Of course, the range of admissible rotation numbers depends on the choice of the lift $\tilde{f}$ of f. If $\tilde{f}$ and $\tilde{f}_1$ are two lifts of f, then they differs by translation by an integer $\tilde{f}_1 = \tilde{f} + (n, 0)$, for some $n \in \mathbf{Z}$. The rotation numbers of the action minimizing sets are changed by the same number, i.e. $M_{\omega+n}^*(\tilde{f}_1) = M_\omega^*(\tilde{f})$. For example, in the case of billiards, the range of admissible rotation numbers may be $n < \omega < n+1$, for any integer n, according to the lift.

As a final example of the application of Theorems 6.1 and 6.2, we consider an elliptic fixed point P of a C^∞ area preserving diffeomorphism f of a surface. According to Birkhoff's normal form theorem, we choose a C^∞ complex local coordinate $\zeta = \xi + i\eta$, centered at P, with respect to which f has the form $f(\zeta) = \zeta'$, where

$$\zeta' = N(\zeta) + O(\rho^k)$$

and $N(\zeta) = \zeta \exp 2\pi i(\beta_0 + \beta_1\rho^2 + ... + \beta_N\rho^{2N})$. We suppose, furthermore, that $d\xi \wedge d\eta$ is the given area form on the surface. We suppose that at least one of the Birkhoff invariants β_1, β_2, ... is non-vanishing. Then, as described in §4, f is locally a twist mapping in a punctured neighborhood of P. We may suppose, without loss of generality, that the first non-vanishing Birkhoff invariant is positive. (Otherwise, replace f by f^{-1}.) Then, in the notation used at the end of §4, $(W, g) \in J^\infty$.

It may be shown that for an appropriate choice of lift $\tilde{g}$ of g, the hypothesis of Theorem 6.2 is satisfied for $\beta_0 < \omega < \beta_0 + \epsilon$, for a sufficiently small positive number ϵ. (cf. §4.) Thus, the action minimizing sets M_ω^* exist for ω in that range.

§7. Properties of the Variational Principle.

In [Au-LeD], Aubry and Le Daeron proved the existence of action minimizing sets for a Chirikov mapping and noted that their method works more generally. Their method differs from the first author's [Ma1] (described in the previous section) and leads to further results. In [Ba], Bangert generalized the Aubry-Le Daeron method, showing that their results still hold under weak conditions on the variation principle. In this section, we discuss Bangert's conditions and some further conditions introduced by the first author [Ma8].

The conditions introduced in this section are satisfied by the variational principle associated to a Chirikov mapping (i.e. the Frenkel-Kontorova model). More generally, they are satisfied for the variational principle associated to any diffeomorphism f of the cylinder such that $(\mathbf{S}^1 \times \mathbf{R}, f) \in J$ and f twists $\mathbf{S}^1 \times \mathbf{R}$ infinitely at each end, provided that an additional uniformity condition explained later is satisfied. In addition, it is possible to associate a variational principle to a finite composition of such mappings in such a way that these conditions are still satisfied. This permits the extension of the theory of action minimizing orbits to such compositions.

Bangert pointed out a further application of his generalization of the Aubry-Le Daeron theory in [Ba]. In 1932, Hedlund [Hd] developed a theory of geodesics on the 2-torus, with an arbitrary C^∞ metric, whose lifts to the universal cover globally minimize arc length. Hedlund called these "class A" geodesics. Bangert showed [Ba] that a large part of the theory of class A geodesics follows from his generalization of the Aubry-Le Daeron theory.

Here are Bangert's conditions on the variational principle h:

(H_0) h is a continuous real valued function whose domain is $\mathbf{R}^2$.

(H_1) $h(x+1, x'+1) = h(x, x')$.

(H_2) $h(x, x') \to +\infty$ as $|x - x'| \to +\infty$.

(H_3) The inequality (5.5) holds.

(H_4) Consider real numbers $\overline{x}$, x, x', $\overline{\xi}$, ξ'. Suppose both of the functions $h(\overline{x}, y) + h(y, x')$ and $h(\overline{\xi}, y) + h(y, \xi')$ of y take a minimum value at $y = x$. Then either $(\overline{x} - \overline{\xi})(x' - \xi') < 0$ or $\overline{x} = \overline{\xi}$ and $x' = \xi'$. For example, if f maps $\mathbf{S}^1 \times \mathbf{R}$ onto itself, $(\mathbf{S}^1 \times \mathbf{R}, f) \in J$, and f twists each end of $\mathbf{S}^1 \times \mathbf{R}$ infinitely, then the variational principle associated to f satisfies $(H_0) - (H_4)$. We have already shown (H_0), (H_1) and (H_3) in §5. Clearly $y, y' \to \pm\infty$ as $x' - x \to \pm\infty$; thus, (H_2) follows from (5.2). Condition (H_4) follows from the facts (§5) that $\partial_1 h(x, x')$ is a strictly decreasing function of x' and $\partial_2 h(x, x')$ is a strictly decreasing function of x. For, if $\overline{\xi} \geq \overline{x}$ and $\xi' \geq x'$ and one of these inequalities is strict, then

$$\frac{d}{dy}\left(h(\overline{\xi}, y) + h(y, \xi')\right) < \frac{d}{dy}\left(h(\overline{x}, y) + h(y, x')\right),$$

which contradicts the assumption that both of the functions $h(\overline{\xi}, y) + h(y, \xi')$ and $h(\overline{x}, y) + h(y, x')$ take a minimum value at $y = x$.

It is often convenient to consider two further conditions, introduced by the first author in [Ma8]. The first of these is a strengthening of (H_3):

(H_5) There exists a positive continuous function ρ on $\mathbf{R}^2$ such that

(7.1)
$$h(\xi, \xi') + h(x, x') - h(\xi, x') - h(x, \xi') \leq -\int_\xi^x \int_{\xi'}^{x'} \rho(u, u') \, du \, du'$$

if $\xi < x$ and $\xi' < x'$.

The second is expressed in terms of a positive number θ.

($H_{6\theta}$) The function
$$\theta (x - x')^2 / 2 - h(x, x')$$
is convex in each variable x and x'.

If h satisfies this condition for some positive number θ, we will say that h satisfies (H_6). If h satisfies (H_1) $-$ (H_5) and ($H_{6\theta}$), we will say that it satisfies (H_θ). If h satisfies (H_θ) for some positive number θ, we will say that it satisfies (H).

The conditions just considered on f are not enough to guarantee that h satisfies (H_5) and (H_6). However, if we impose, in addition to the other conditions, the condition that $(\mathbf{S}^1 \times \mathbf{R}, f) \in J^1$ (i.e. a differentiability hypothesis), then h satisfies (H_5), by (5.4). Indeed, we may set $\rho = -\partial_{12} h$ and then we have equality in (7.1).

For (H_6) to hold, we need a further strengthening of the differentiability hypothesis. If we set $f(x, y) = (x', y')$, the condition that $(\mathbf{S}^1 \times \mathbf{R}, f) \in J^1$ may be expressed as two conditions: f is a C^1 diffeomorphism and $\partial x'/\partial y > 0$. Also, this condition may be expressed in terms of f^{-1}, namely f^{-1} is a C^1 diffeomorphism and $\partial x/\partial y' < 0$, where x' and y' are taken as independent variables.

If, in addition to the conditions we have already imposed on f, we impose

(7.2)
$$\frac{\partial y'/\partial y}{\partial x'/\partial y} \leq \theta \quad \text{and} \quad \frac{\partial y/\partial y'}{\partial x/\partial y'} \geq -\theta \,,$$

then the associated variational principle satisfies ($H_{6\theta}$). (In the first of these inequalities, we take x and y as independent variables and x' and y' as dependent variables; in the second, we take x' and y' as independent variables and x and y as dependent variables).

Indeed, in this case h is C^2, and

(7.3)
$$\partial_{22} h(x, x') = \frac{\partial y'/\partial y}{\partial x'/\partial y} \quad \text{and} \quad \partial_{11} h = -\frac{\partial y/\partial y'}{\partial x/\partial y'} \,.$$

These equations follow from (5.2) by a calculus exercise. In doing the exercise, it is important to keep in mind that x and x' are the independent variables on the left side of each of these equations, whereas x and y are the independent variables on the right side of the first of these equations and x' and y' are the independent variables on the right side of the second of these equations.

From (7.3), it is clear that

(7.4)
$$\partial_{11} h(x, x') \leq \theta \quad \text{and} \quad \partial_{22} h(x, x') \leq \theta \,.$$

These two conditions are equivalent to ($H_{6\theta}$) in the case that h is C^2.

The conditions (7.2) may be interpreted geometrically. The positive monotone twist condition implies that f turns a vertical vector to the right and f^{-1} turns a vertical vector to the left. Let $\theta = \cot \beta$, where $0 < \beta < \pi/2$. The first inequality in (7.2) is equivalent to the condition that f turns every vertical vector to the right by an angle at least β. The second inequality in (7.2) is equivalent to the condition that f^{-1} turns every vertical vector to the left by an angle of at least β. The second inequality in (7.2) is equivalent to the conditions that f^{-1} turns every vertical vector to the left by an angle of at least β.

This leads us to introduce the following notation: for $r \geq 1$, we denote by J_θ^r the set of diffeomorphisms of $\mathbf{S}^1 \times \mathbf{R}$ onto itself such that $(\mathbf{S}^1 \times \mathbf{R}, f) \in J^r$, and f turns every vertical vector to the right by an angle of at least β, and f^{-1} turns every vertical vector to the left by at least β, where $\beta = \cot^{-1} \theta$, $0 < \beta < \pi/2$. We let $J_*^r = \cup_{\theta > 0} J_\theta^r$. We may summarize the discussion above as:

Theorem 7.1. If $f \in J_\theta^r$, then its associated variational principle h satisfies (H_θ). If $f \in J_*^r$, then h satisfies (H).

It is clear that a Chirikov mapping is a member of J_*^∞. On the other hand, the condition (H) is not satisfied by the variational principle associated to a billiards mapping, nor by the variational principle associated to a neighborhood of an elliptic fixed point of an area-preserving mapping with non-vanishing Birkhoff invariant. Nonetheless, the extension theorem of the next section allows us to apply results about variational principles satisfying (H) to these mappings.

§8. An Extension Theorem.

Theorem 8.1. *Let* β_-, $\beta_+ : \mathbf{R} \to \mathbf{R}$ *be* C^{r-1} *diffeomorphisms satisfying* $\beta_\pm(x + 1) = \beta_\pm(x) + 1$, *where* $r \geq 1$. *Suppose furthermore that* $\beta_-(x) < \beta_+(x)$. *Let* $W = \{(x, x') \mid \beta_-(x) \leq x' \leq \beta_+(x)\}$. *Let* $h : W \to \mathbf{R}$ *be a* C^{r+1} *function such that* $h(x+1, x'+1) = h(x, x')$ *and* $\partial_{12} h < 0$. *Then* h *extends to a* C^{r+1} *function defined on all of* $\mathbf{R}^2$ *satisfying* (H).

Proof. W is invariant under the translation $(x, x') \to (x+1, x'+1)$. Its quotient by this translation is compact. Consequently, there exists a number $\delta > 0$ such that $\partial_{12} h \leq -\delta$. It is well known that a C^{r-1} function on a C^{r-1} submanifold extends to the ambient space. Applied here, this implies that there exists a C^{r-1} function ρ on $\mathbf{R}^2$ such that $\rho(x + 1, x' + 1) = \rho(x, x')$, $\rho \leq -\delta$ and $\rho|W = -\partial_{12} h$. In addition, we may suppose $\rho = $ constant outside of the set $\{\beta_-(x) - 1 \leq x' \leq \beta_+(x) + 1\}$. Then there is a unique extension of h to all of $\mathbf{R}^2$ satisfying $\rho = \partial_{12} h$. This may be seen as follows: The value of $\partial_1 h$ on graph β_- and the relation $\partial_2(\partial_1 h) = \rho$ determine $\partial_1 h$ on all of $\mathbf{R}^2$. Then the value of h on graph β_- and $\partial_1 h$ determine h on all of $\mathbf{R}^2$.

From this procedure for determining h, the periodicity condition (H_1) is obvious. Since $\partial_2(\partial_1 h) = \rho \leq -\delta$, it follows that $\partial_1 h(x, x') \to \pm\infty$ as $x - x' \to \pm\infty$. (H_2) follows. (H_3) and the stronger condition (H_5) follow immediately from $\partial_{12} h \leq -\delta$, as does (H_4) by an argument already discussed. Finally (H_6) holds in $\{(x, x') \mid \beta_-(x) - 1 \leq x' \leq \beta_+(x) + 1\}$ by compactness. Since $\partial_2(\partial_{11} h) = \partial_1(\partial_{12} h) \equiv 0$ outside $\{(x, x') \mid \beta_-(x) - 1 \leq x' \leq \beta_+(x) + 1\}$, a bound $\partial_{11} h \leq \theta$ which holds in $\{(x, x') \mid \beta_-(x) - 1 \leq x' \leq \beta_+(x) + 1\}$ holds in all of $\mathbf{R}^2$. □

This extension theorem may be applied, for example, to the billiard mapping $f : U \to \mathbf{S}^1 \times \mathbf{R}$, as follows. Let $h : V \to \mathbf{R}$ be the associated variational principle, where $V = \{(x, x') \mid x < x' < x + 1\}$. Let $\delta > 0$ be a small positive number. Let $W_\delta = \{(x, x') \mid x + \delta \leq x' \leq x + 1 - \delta\}$. Then $h|W_\delta$ extends to all of $\mathbf{R}^2$. Any orbit of the billiards which is bounded away from the boundary corresponds to a stationary configuration of $h|W_\delta$ for some $\delta > 0$.

In a similar way, if P is an elliptic fixed point of an area preserving mapping, and one of the Birkhoff invariants is non-zero, then we can reduce the study of any orbit which lies in a sufficiently small neighborhood of P, and at the same time is bounded away from P, to the study of the corresponding stationary configuration of a suitably chosen variational principle (constructed by means of the extension theorem) which satisfies condition (H).

§9. Minimal Configurations.

We have already briefly mentioned the notion of minimal configuration in §3 in our discussion of the Frenkel-Kontorova model. In this section, we will define the notion of a minimal configuration for a general variational principle. The definition we give in this section is different from that given in §3. However, we will show that in the case that h satisfies (H), the two definitions are equivalent. This applies in particular to the Frenkel-Kontorova model.

By a *configuration*, we will mean a bi-infinite sequence $x = (..., x_i, ...) \in \mathbf{R}^{\mathbf{Z}}$ of real numbers. By a *segment* of a configuration, we will mean a finite sequence $x = (x_k, ..., x_l)$. If $h : \mathbf{R}^2 \to \mathbf{R}$, we set

$$h(x) = h(x_k, ..., x_l) = \sum_{i=k}^{l-1} h(x_i, x_{i+1}) \ .$$

A segment will be said to be *h-minimal* if, for every segment $(\xi_k, ..., \xi_l)$ such that $x_k = \xi_k$ and $x_l = \xi_l$ (but ξ_i can be arbitrary for $k < i < l$), we have

$$h(x_k, ..., x_l) \le h(\xi_k, ..., \xi_l) \ .$$

A configuration will be said to be *h-minimal* if each segment of it is h-minimal.

In [Au-LeD], Aubry and Le Daeron developed a theory of h-minimal configurations for the Frenkel-Kontorova model. Bangert [Ba] generalized their theory to a variational principle satisfying $(H_0) - (H_4)$. In the following sections we briefly sketch the Aubry-Le Daeron-Bangert theory.

Definition. Let $x = (x_k, ..., x_l)$ be a segment of a configuration. By the *Aubry graph* $G(x)$ of such a segment, we will mean the union of the line segments in $\mathbf{R}^2$ joining (i, x_i) and $(i + 1, x_{i+1})$, for $k \le i < l$.

Aubry's Crossing Lemma. Let $x = (x_k, ..., x_l)$ and $y = (y_k, ..., y_l)$ denote two h-minimal configurations. Then $G(x) \cap G(y)$ contains at most two points. If it contains two points, then these are the endpoints of the two graphs, i.e. $x_k = y_k$ and $x_l = y_l$.

We introduce the following notation:

$$(x \vee y)_i = \max(x_i, y_i) \quad \text{and} \quad (x \wedge y)_i = \min(x_i, y_i) \ .$$

The proof of Aubry's crossing Lemma is based on the following inequality

(9.1) $$h(x \vee y) + h(x \wedge y) - h(x) - h(y) \le 0 \ .$$

This inequality follows immediately from (H_3). Furthermore, it is clear from (H_3) that if $G(x)$ crosses $G(y)$ in the interior of the interval, then the inequality (9.1) is strict.

Proof of Aubry's Crossing Lemma. Assume that the conclusion is false. By replacing x and y with subsegments and interchanging the roles of x and y if necessary, we may assume without loss of generality that $x_k \le y_k$ and $x_l \le y_l$, and at least one of these inequalities is strict. From the assumption that x (resp. y)

is h-minimal, it follows that $h(x) \leq h(x \wedge y)$ (resp. $h(y) \leq h(x \vee y)$). Combining this with (9.1), we see that these are equalities and that (9.1) is an equality. Since (9.1) is an equality, $G(x)$ cannot cross $G(y)$ in the interior of an interval. Since $h(x) = h(x \wedge y)$, we have that $x \wedge y$ is h-minimal. Since $h(y) = h(x \vee y)$, we have that $x \vee y$ is h-minimal.

Since $G(x)$ and $G(y)$ do not cross in the interior of an interval, but they do intersect at other than an endpoint, we have that $G(x \vee y)$ meets $G(x \wedge y)$ at a node which is not an endpoint. This contradicts (H_4). $\qquad\square$

§10. Existence of Periodic Minimal Configurations.

We say that a configuration $x = (..., x_i, ...)$ is *periodic of type* (p, q) if $x_{i+q} = x_i + p$. We set

$$h_q^{per} = \sum_{i=a}^{a+q-1} h(x_i, x_{i+1}) \; .$$

This is obviously independent of the choice of a. We assume that h satisfies $(H_0) - (H_4)$.

Theorem 10.1. h_q^{per} *takes a minimum value on the set of configurations which are periodic of type* (p, q).

Proof. We may identify the set of configurations of type (p, q) with $\mathbf{R}^q$. In view of (H_1), h_q^{per} is invariant under the translation $T(..., x_i, ...) = (..., x_i + 1, ...)$. Thus h_q^{per} is defined on the quotient space $\mathbf{R}^q/T$. In vie of (H_2), $\{h_q^{per} \le a\}$ is a compact subset of $\mathbf{R}^q/T$, for any $a \in \mathbf{R}$. Since h_q^{per} is continuous, it takes a minimum value. $\square$

Theorem 10.2. *If* $p, q, n \in \mathbf{Z}$, $q \ne 0$, $n \ne 0$, *and* $x = (..., x_i, ...)$ *minimizes* h_q^{per} *over the set of periodic configurations of type* (pn, qn), *then* x *is periodic of type* (p, q).

Proof. Let $y_i = x_{i+q} - p$. Clearly x is periodic of type (p, q) if and only if $x = y$. Suppose $x \ne y$. It is easily seen that the Aubry graphs of x and y meet infinitely often. An argument like the proof of Aubry's Crossing Lemma shows that this contradicts the fact that both x and y minimize h_{qn}^{per} over the set of periodic configurations of type (pn, qn). $\square$

Theorem 10.3. *If* x *is a periodic configuration of type* (p, q) *which minimizes* h_q^{per} *over all such configurations, then* x *minimizes* h_{qn}^{per} *over all periodic configurations of type* (pn, qn). *Moreover,* x *is minimal in the sense of* §9.

Proof. By Theorem 10.1, there exists a configuration y which minimizes h_{qn}^{per} over all periodic configurations of type (pn, qn). By Theorem 10.2, y is periodic of type (p, q). Clearly, y minimizes h_q^{per} over all periodic configurations of type (p, q). Hence $h_q^{per}(x) = h_q^{per}(y)$ and $h_{qn}^{per}(x) = n\, h_q^{per}(x) = h_{qn}^{per}(y)$. Since y minimizes and x takes the same value, x minimizes. This proves the first statement.

To prove the second statement, i.e. a segment $(x_k, ..., x_l)$ of x minimizes $h(x_k, ..., x_l)$ subject to the fixed boundary condition, it is enough to apply the first statement with $qn > l - k + 1$. $\square$

Corollary 10.4. *There exists an* h-*minimal configuration which is periodic of type* (p, q).

§11. The Rotation Number.

Let x and y be configurations. We define the relations $>_\alpha$, $>_\omega$, $<_\alpha$ and $<_\omega$ as follows: $x >_\alpha y$ (resp. $x >_\omega y$, $x <_\alpha y$, $x <_\omega y$) if there exists i_0 such that $x_i > y_i$ for all $i < i_0$ (resp. $x_i > y_i$ for all $i > i_0$, etc.). As before, we suppose that h satisfies $(H_0) - (H_4)$. We have

Trichotomy I. *if x and y are h-minimal, then $x >_\alpha y$, $x = y$ or $x <_\alpha y$ and $x >_\omega y$, $x = y$ or $x <_\omega y$.*

This follows immediately from Aubry's Crossing Lemma, which implies that the Aubry graphs of x and y cross at most once.

If x is a configuration and p, $q \in \mathbf{Z}$, we define the *translate* $(T_{p,q} x)_i = x_{i-q} + p$. Obviously, if $T_{p,q} x >_\alpha x$, then $T_{pl,ql} x >_\alpha x$ for all positive integers l. By Trichotomy I, we may say "if and only if", in the case that x is h-minimal. The same comments apply to the relations $>_\alpha$, $>_\omega$, $<_\alpha$ and $<_\omega$. Thus, we may define

$$A_\alpha(x) = \{p/q \mid T_{p,q} x <_\alpha x , \quad p, q \in \mathbf{Z}, q > 0\}$$
$$B_\alpha(x) = \{p/q \mid T_{p,q} x >_\alpha x , \quad p, q \in \mathbf{Z}, q > 0\}$$

and similarly $A_\omega(x)$ and $B_\omega(x)$ (replace $>_\alpha$ and $<_\alpha$ by $>_\omega$ and $<_\omega$).

It is easy to see that if x is h-minimal and p/q, p'/q' are rational numbers with $p'/q' < p/q$, then $p/q \in A_\alpha(x)$ implies $p'/q' \in A_\alpha(x)$ and $p'/q' \in B_\alpha(x)$ implies $p/q \in B_\alpha(x)$. The same result holds for α replaced by ω. These observations follow from the fact that $(T_{p'q,q'q} x)_i < (T_{pq',qq'} x)_i$, since $p'q < pq'$. Moreover, when x is h-minimal, $A_\alpha(x) \cup B_\alpha(x)$ (resp. $A_\omega(x) \cup B_\omega(x)$) is all of $\mathbf{Q}$ except possibly one element. Consequently, there exists a unique $\rho_\alpha(x)$ (resp. $\rho_\omega(x)$) $\in \{-\infty\} \cup \mathbf{R} \cup \{+\infty\}$ such that $p/q < \rho_\alpha(x)$ (resp. $\rho_\omega(x)$) implies that $p/q \in A_\alpha(x)$ (resp. $A_\omega(x)$) and $p/q > \rho_\alpha(x)$ (resp. $\rho_\omega(x)$) implies $p/q \in B_\alpha(x)$ (resp. $B_\omega(x)$).

Theorem 11.1. $\rho_\alpha(x) = \rho_\omega(x)$, *if x is h-minimal.*

Proof. Suppose the contrary, e.g. $\rho_\alpha(x) < \rho_\omega(x)$. Let p/q be a rational number, expressed in lowest terms, with $q > 0$, such that $\rho_\alpha(x) < p/q < \rho_\omega(x)$. By Corollary 10.4, there exists an h-minimal configuration y which is periodic of type (p, q). From the inequalities $\rho_\alpha(x) < p/q < \rho_\omega(x)$, it follows that $x >_\alpha y$ and $x >_\omega y$, and this holds for any translate of y. However, if r is sufficiently large, then $x_i \geq (T_{r,1} y)_i$ does not hold for all i. Hence, the Aubry graphs of x and y cross at least twice. Since x and y are h-minimal, this contradicts Aubry's Crossing Lemma.

This contradiction shows that we do not have $\rho_\alpha(x) < \rho_\omega(x)$. Similarly, we do not have $\rho_\omega(x) < \rho_\alpha(x)$. $\qquad\square$

We set $\rho(x) = \rho_\alpha(x) = \rho_\omega(x)$.

Theorem 11.2. $\rho(x) \in \mathbf{R}$, if x is h-minimal.

Proof. Otherwise, we would have $\rho(x) = \pm\infty$. Suppose, for example, that $\rho(x) = +\infty$. If y is a periodic configuration, $y >_\alpha x$ and $y <_\omega x$. By Corollary 10.4, there exists a periodic minimal configuration y of type $(p, 1)$, for any integer p. Thus, $y_{i+1} = y_i + p$. By choosing p very large, and replacing y by a translate, if necessary, we may suppose that $y_0 < x_0$ and $y_1 > x_1$. The relations $y >_\alpha x$, $y <_\omega x$, $y_0 < x_0$ and $y_1 > x_1$ imply that the Aubry graphs of x and y cross at least three times. This contradicts the Aubry's Crossing Lemma.

This contradiction shows that $\rho(x) \neq +\infty$. Similarly, it may be shown that $\rho(x) \neq -\infty$. $\qquad\square$

We will call $\rho(x)$ the *rotation number* of x.

We let $\mathcal{M} = \mathcal{M}_h \subset \mathbf{R}^\infty$ denote the set of all h-minimal configurations. We provide $\mathbf{R}^\infty$ with the product topology and $\mathcal{M}$ with the induced topology.

Theorem 11.3. $\mathcal{M}$ is closed in $\mathbf{R}^\infty$ and $\rho : \mathcal{M} \to \mathbf{R}$ is continuous.

Proof. The first assertion follows immediately from the definition of $\mathcal{M}$ and the assumption (H_0) that h is continuous. To prove the second, we will use:

Lemma. Let $x = (..., x_i, ...) \in \mathcal{M}$. Let $p, q \in \mathbf{Z}$, $q \geq 1$. If $x_{i+q} \leq x_i + p$, for some i, then $\rho(x) \leq (p+1)/q$. If $x_{i+q} \geq x_i + p$, for some i, then $\rho(x) \geq (p-1)/q$.

Proof. To prove the first assertion, we use the existence of a periodic minimal configuration y of type $(p + 1, q)$. By replacing y with a suitable translate, if necessary, we may suppose that $y_i < x_i$ and $y_{i+q} > x_i + p \geq x_{i+q}$. By the Aubry's Crossing Lemma $y <_\alpha x$ and $y >_\omega x$, so $\rho(x) \leq \rho(y) = (p+1)/q$. The second assertion may be proved similarly. $\qquad\square$

Now the continuity of ρ follows easily, since x_{i+q} is obviously a continuous function of x. $\qquad\square$

We let $\mathrm{pr}_i : \mathbf{R}^\infty \to \mathbf{R}$ denote the projection on the ith factor.

Theorem 11.4. Let I, Ω be compact subsets of $\mathbf{R}$. For any $i \in \mathbf{Z}$, $\rho^{-1}(\Omega) \cap \mathrm{pr}_i^{-1}(I)$ is a compact subset of $\mathcal{M}$.

Proof. By the Tychonoff product theorem and the fact that $\mathcal{M}$ is closed in $\mathbf{R}^\infty$, it is enough to show that for each $j \in \mathbf{Z}$, $\mathrm{pr}_j(\rho^{-1}(\Omega) \cap \mathrm{pr}_i^{-1}(I))$ is a bounded subset of $\mathbf{R}$. In the case that $j > i$ an upper bound for this set may be found as follows. Let $p \in \mathbf{Z}$ be an upper bound for Ω and let y be a periodic minimal configuration of type $(p, 1)$ such that y_i is an upper bound for I. Then y_j is an upper bound for $\mathrm{pr}_j(\rho^{-1}(\Omega) \cap \mathrm{pr}_i^{-1}(I))$. For if $x \in \rho^{-1}(\Omega) \cap \mathrm{pr}_i^{-1}(I)$, then $y >_\omega x$ since $\rho(y) = p > \rho(x) \in \Omega$. Moreover, $y_i > x_i$, so $y_j > x_j$ by the Aubry's Crossing Lemma.

The other cases may be treated similarly. $\qquad\square$

Theorem 11.5. *Let I be a closed interval of unit length in $\mathbf{R}$. For any $i \in \mathbf{Z}$, $\rho : \mathrm{pr}_i^{-1}(I) \cap \mathcal{M} \to \mathbf{R}$ is surjective.*

Proof. For any rational number p/q, there exists a minimal configuration $x = (\ldots, x_i, \ldots)$ of type (p, q) by Corollary 10.4. Since I has unit length, we may suppose that $x_i \in I$. Since $\rho(x) = p/q$, we obtain that $p/q \in \rho(\mathrm{pr}_i^{-1}(I) \cap \mathcal{M})$. Thus $\rho(\mathrm{pr}_i^{-1}(I) \cap \mathcal{M})$ contains $\mathbf{Q}$. On the other hand, by Theorem 11.4, if $\Omega \subset \mathbf{R}$ is compact then

$$\rho(\mathrm{pr}_i^{-1}(I) \cap \mathcal{M}) \cap \Omega = \rho(\mathrm{pr}_i^{-1}(I) \cap \rho^{-1}(\Omega))$$

is compact. It follows that $\rho(\mathrm{pr}_i^{-1}(I) \cap \mathcal{M})$ is closed. Since this set is closed and contains $\mathbf{Q}$, it contains all of $\mathbf{R}$. $\qquad\square$

§12. Irrational Rotation Number.

Let ω be an irrational rotation number. Let h be a variational principle which satisfies the Bangert conditions $(H_0)-(H_4)$. By Theorem 11.5, there exist h-minimal configurations x of rotation number ω. We let $\mathcal{M}_\omega = \mathcal{M}_{h,\omega}$ denote the set of h-minimal configurations of rotation number ω. By Theorem 11.3, $\mathcal{M}_\omega$ is closed in $\mathbf{R}^\infty$. In this section, we will describe several results concerning the structure of $\mathcal{M}_\omega$, due to Aubry and Le Daeron [Au-LeD] and later in a more general context to Bangert [Ba].

The first result concerns the order relation. Given two configurations x and y, we will say that $x < y$ (or $y > x$) if $x_i < y_i$, for every i. If x and y are h-minimal configurations, then $x < y$ if and only if $x <_\alpha y$ and $x <_\omega y$. This follows from Aubry's Crossing Lemma.

Trichotomy II. *If x and y are h-minimal configurations and have the same irrational rotation number, then $x < y$, $x = y$ or $x > y$.*

We will follow the traditional approach ([Au-LeD], [Ba]) and prove this in several steps. For the first step, we prove this when y is a translate of x, say $y = T_{p,q}\, x$. Since $\rho_\alpha(x) = \rho_\omega(x)$ and this number is irrational, $A_\alpha(x) = A_\omega(x)$ and $B_\alpha(x) = B_\omega(x)$. This is because when $\rho_\alpha(x)$ is irrational $A_\alpha(x) = \{p/q < \rho_\alpha(x)\}$ and $B_\alpha(x) = \{p/q > \rho_\alpha(x)\}$ (and similarly for ω in place of α). From the definition of $A_\alpha(x)$ etc., it follows that $y <_\alpha x$ (resp. $y >_\alpha x$) if and only if $y <_\omega x$ (resp. $y >_\omega x$). Then the result follows from Trichotomy I.

The second step is the definition of what Aubry called the *hull function* of a minimal configuration x of irrational rotation number ω. Let

$$\phi_x(t) = \sup\{(T_{p,q}\, x)_0 \mid p - q\omega \le t\} \ .$$

Thus ϕ is a mapping of $\mathbf{R}$ into $\mathbf{R}$. Note that $T_{p,q}\, x < T_{p',q'}\, x$ if and only if $p - q\omega < p' - q'\omega$. (This is a consequence of the trichotomy property which we just proved, and the definition of the rotation number.) Consequently, using the assumption that ω is irrational, we see that ϕ_x is strictly increasing, i.e.

$$s < t \Rightarrow \phi_x(s) < \phi_x(t) \ .$$

Furthermore $\phi_x(t+1) = \phi_x(t) + 1$, as may be seen from the definition of ϕ_x.

Aubry called ϕ_x the *hull function* of x.

Using Aubry's hull function, we may define new minimal configurations as follows: if $t \in \mathbf{R}$, we set

$$x_i^{t\pm 0} = \phi_x(t + \omega i \pm 0) \ .$$

Then $x^{t\pm 0} = (..., x_i^{t\pm 0}, ...)$ is a minimal configuration of rotation number ω. For, it is possible to choose a sequence of pairs (p_i, q_i) such that $p_i - q_i\omega \nearrow t$. Then $x^{t-0} = \lim T_{p_i,q_i}\, x$ and the fact that x^{t-0} is a minimal configuration follows from the fact that $\mathcal{M}_\omega$ is closed. The case of x^{t+0} may be treated similarly.

The third step is:

Theorem 12.1. *If x and y are minimal configurations having the same rotation number, then there exists a constant a such that*

$$\phi_y(t \pm 0) = \phi_x(t + a \pm 0) \,,$$

for all $t \in \mathbf{R}$.

Proof. Let $I(\phi_x)$ denote the unique non-decreasing function of $\mathbf{R}$ into itself which is the inverse of ϕ_x. Theorem 12.1 is equivalent to the statement that $I(\phi_x)$ and $I(\phi_y)$ differ by a constant. Suppose otherwise. Choose a number a such that

$$\min(I(\phi_y) - I(\phi_x)) < a < \max(I(\phi_y) - I(\phi_x)) \,.$$

It is easy to see that for any $t \in \mathbf{R}$, the Aubry graphs of $y^{t \pm 0}$ and $x^{t+a \pm 0}$ cross infinitely often. But, since these are minimal configurations, this provides a contradiction to the Aubry Crossing Lemma.

Clearly, if $a < 0$ (resp.> 0), then $x < y$ (resp. $x > y$). Thus, in order to prove Trichotomy II, it is enough to consider the case $a = 0$. In this case $x^{t \pm 0} = y^{t \pm 0}$ for all t. We use the abbreviation $x^{\pm 0}$ for $x^{0 \pm 0}$. Thus,

(12.1) $$x^{-0} < x < x^{+0} \quad \text{and} \quad x^{-0} < y < x^{+0} \,.$$

From the fact that ϕ_x is strictly increasing and the periodicity condition $\phi_x(t+1) = \phi_x(t) + 1$, together with the assumption that ω is irrational, it follows that

(12.2) $$\sum_{-\infty}^{\infty} x_i^+ - x_i^- \leq 1 \,.$$

From (12.1) and (12.2), it follows that x and y are *asymptotic*, i.e. $|x_i - y_i| \to 0$, as $i \to \pm\infty$. More generally, we will say that x and y are $\alpha - asymptotic$ if $|x_i - y_i| \to 0$, as $i \to -\infty$ and $\omega - asymptotic$ if $|x_i - y_i| \to 0$, as $i \to +\infty$. The fourth step in the proof of Trichotomy II is the following:

Addendum to Aubry's Crossing Lemma. *If x and y are h-minimal configurations which are $\alpha-$ or $\omega-$asymptotic, then $x = y$ or their Aubry graphs do not meet.*

Proof. Suppose, for example, that x and y are $\omega-$asymptotic and their graphs meet. Choose integers $k < l$ such that their graphs meet in the interval between k and l. Let $x_i^\star = x_i$, for $i \neq l$ and $x_l^\star = y_l$. The argument used to prove Aubry's Crossing Lemma shows that there exists segments $(\xi_k, ..., \xi_l)$ and $(\eta_k, ..., \eta_l)$ of configurations such that $\xi_k = x_k^\star$, $\eta_k = y_k$, $\xi_l = \eta_l = x_l^\star = y_l$, and

$$h(\xi_k, ..., \xi_l) + h(\eta_k, ..., \eta_l) \leq h(x_k^\star, ..., x_l^\star) + h(y_k, ..., y_l) - \epsilon \,.$$

Moreover, ϵ is independent of l, since the construction used in the proof of Aubry's Crossing Lemma is made in the neighborhood of the point where the graphs intersect and is unaffected by anything not near this point.

Thus, it is enough to show that

(12.3) $$h(x_k^\star, ..., x_l^\star) \leq h(x_k, ..., x_l) + \epsilon \,,$$

if l is large enough. But

$$(12.4) \qquad \rho(x) - 2 \le x_{i+1} - x_i \le \rho(x) + 2 \; ,$$

by the Lemma used in the proof of Theorem 11.3, and $h(x_i, x_{i+1})$ is uniformly continuous in the region defined by (12.4), since it is continuous and satisfies the periodicity condition $h(x + 1, x' + 1) = h(x, x')$. Therefore (12.3) holds for large l, since x and y are ω−asymptotic. $\square$

Now Trichotomy II follows from the addendum, because we have already reduced the proof to the case when x and y are asymptotic. Since the graphs of x and y do not cross, if $x \ne y$ we have $x < y$ or $x > y$. $\square$

We set $A_\omega = A_\omega^h = \mathrm{pr}_0(\mathcal{M}_\omega^h)$. Trichotomy II has the following important consequence:

Theorem 12.2. A_ω^h *is a closed subset of* $\mathbf{R}$ *and* $\mathrm{pr}_0 : \mathcal{M}_\omega^h \to A_\omega^h$ *is a homeomorphism for any irrational number* ω.

Proof. By Trichotomy II, this mapping is injective. By Theorem 11.4, it is proper. The two assertion then follow from elementary results in general topology. $\square$

We define two homeomorphisms T (translation) and S (shift) of $\mathcal{M}_\omega^h$ onto itself, as follows:

$$T(..., x_i, ...) = (..., x_i + 1, ...) \quad \text{and} \quad S(..., x_i, ...) = (..., x_{i+1}, ...) \; .$$

Obviously these commute, so S induces a homeomorphism $\overline{S}$ of $\mathcal{M}_\omega^h/T$ onto itself.

Obviously, $\mathcal{M}_\omega^h/T$ is compact. Recall that if f is a homeomorphism of a compact topological space, a closed invariant subset K in X is said to be *minimal* if it contains no closed invariant sets other that itself and the empty set. Birkhoff showed that every homeomorphism of a compact space has a minimal set. (This follows immediately from Zorn's Lemma). Denjoy [De] showed that every orientation preserving homeomorphism f of the circle of irrational rotation number has exactly one minimal set, and every orbit tends to it under forward and backward iteration. Moreover, he showed that such an f is semi-conjugate to the rotation R_ω of the same rotation number, i.e. there exists $\Phi : \mathbf{S}^1 \to \mathbf{S}^1$ such that $\Phi \circ f = R_\omega \circ \Phi$. Moreover, for each $\theta \in \mathbf{S}^1$, $\Phi^{-1}(\theta)$ is one point or an interval.

Theorem 12.3. $\overline{S} : \mathcal{M}_\omega^h/T \to \mathcal{M}_\omega^h/T$ *has a unique minimal set, to which every orbit tends under forward and backward iteration.* $\overline{S}$ *is semi-conjugate to the rotation* R_ω.

Proof. This is reinterpretation of what we already proved. If x is a minimal configuration, the minimal set is $\{x^{t\pm 0} \,|\, t \in \mathbf{R}\}$. This set is minimal because every orbit of $\overline{S}$ is dense in it. It is the only minimal set, because every orbit of $\overline{S}$ tends to it.

The generalized inverse of ϕ_x provides a semiconjugacy to a rotation. $\square$

§13. Rational Rotation Number.

Let $\omega = p/q$, $p, q \in \mathbf{Z}$, $q \geq 1$, $(p, q)=1$. Let h be a variational principle which satisfies the Bangert conditions $(H_0) - (H_4)$. In this section, we continue to describe results of Aubry and Le Daeron [Au-LeD], generalized by Bangert [Ba], this time concerning configurations of rational rotation number.

Let x be a minimal configuration of rotation number p/q. We first consider the case when $T_{-p,-q}\, x >_\omega x$. Then $x_{i+ql} - pl$, $l = 1, 2, \dots$ is an eventually increasing sequence. Moreover, $x_{i+ql} - pl \leq x_i + 2$, since we would have $\rho(x) \geq (pl + 1)/ql$ by the Lemma used in the proof of Theorem 11.3. Therefore $\lim_{l \to +\infty} x_{i+ql} - pl$ exists. We denote the limit by $(l_\omega\, x)_i$. Then $l_\omega x$ is a minimal configuration and

$$(13.1) \qquad l_\omega x = \lim_{l \to +\infty} T_{-pl,-ql}\, x \; .$$

In the case that $T_{-q,-p}\, x <_\omega x$ a similar argument shows that (13.1) exists. Similarly

$$(13.2) \qquad l_\alpha x = \lim_{l \to -\infty} T_{-pl,-ql}\, x$$

exists. Obviously, both $l_\alpha x$ and $l_\omega x$ are periodic of type (p, q).

If x is periodic of type (p, q) then $T_{-p,-q}\, x = x$, so $l_\alpha x = l_\omega x = x$. Otherwise, the Aubry graphs of x, $l_\alpha x$ and $l_\omega x$ do not meet by the Addendum to Aubry's Crossing Lemma (§12). Moreover, the Aubry graphs of $T_{-p,-q}\, x$ and x do not meet, by the same Addendum, since obviously $l_\alpha T_{-p,-q}\, x = l_\alpha x$ and $l_\omega T_{-p,-q}\, x = l_\omega x$. From these observations, we may draw two consequences.

Trichotomy III. *If x and y are minimal configurations which are translates of one another, then their Aubry graphs do not cross, unless $x = y$.*

Proof. We have already shown this in the case that the rotation number of x (and y) is irrational (see the beginning of the proof of Trichotomy II in §12). The argument given in the case ω was irrational shows in the present case that the graphs of $T_{r,s}\, x$ and x do not meet if $r/s \neq p/q$. Thus we are left to consider whether the graphs of x and $T_{p,q}\, x$ meet. But, we have just shown that they do not meet unless $x = T_{p,q}\, x$. $\qquad \square$

Thus, we have three alternatives: $T_{-p,-q}\, x < x$, $T_{-p,-q}\, x = x$, or $T_{-p,-q}\, x > x$. We say that x has *rotation symbol* $p/q-$ in the first case, p/q in the second case and $p/q+$ in the third. Clearly, $l_\omega x < x < l_\alpha x$, when x has rotation symbol $p/q-$, $l_\omega x = x = l_\alpha x$ when x has rotation symbol p/q, and $l_\alpha x < x < l_\omega x$ when x has rotation symbol $p/q+$.

We will use the following notation: we let $\mathcal{M}_{p/q}$ denote the set of periodic minimal configurations of type (p, q) (i.e. of rotation symbol p/q), and we let $\mathcal{M}_{p/q+}$ (resp. $\mathcal{M}_{p/q-}$) denote the union of the set of minimal configurations of rotation symbol $p/q+$ (resp. $p/q-$) and $\mathcal{M}_{p/q}$. It is easy to see that each of these sets is a closed subset of $\mathcal{M}$.

For the study of minimal configurations of type $p/q\pm$, the following notation will be useful: if x and y are two configurations, we set

$$(13.3) \qquad \Delta h(y,x) = \sum_{-\infty}^{\infty}(h(y_i, y_{i+1}) - h(x_i, x_{i+1})) ,$$

provided that the sum is convergent, i.e.

$$\lim_{M\to-\infty,\, N\to+\infty} \sum_{M}^{N}(h(y_i, y_{i+1}) - h(x_i, x_{i+1}))$$

exists. In this case, we will say that $\Delta h(y,x)$ exists. We will also say that $\Delta h(y,x)$ exists when the sum converges to $-\infty$ or to $+\infty$.

Theorem 13.1. If x is a minimal configuration and y is a configuration such that $|x_i - y_i| \to 0$ as $i \to \pm\infty$, then $\Delta h(y,x)$ exists and $\Delta h(y,x) \geq 0$, although $\Delta h(y,x) = +\infty$ is not excluded.

Proof. Since x is minimal (12.4) holds. Moreover, h is uniformly continuous in the region defined by (12.4), by its continuity and periodicity. Let $y^{M,N}$ denote the segment of a configuration, defined for $M \leq i \leq N$ by $y_i^{M,N} = x_i$, for $i = M, N$ and $y_i^{M,N} = y_i$, for $M < i < N$. Since h is uniformly continuous in the region defined by (12.4), it follows that

$$|h(y^{M,N}) - h(y_M, ..., y_N)| \to 0 , \quad \text{as } |M|, |N| \to +\infty ,$$

since $|x_i - y_i| \to 0$, as $i \to \pm\infty$. Since x is minimal,

$$h(y^{M,N}) \geq h(x_M, ..., x_N) .$$

Consequently, for every $\epsilon > 0$, there exists a natural number N, such that if $|M|, |N| \geq N_0$, then

$$h(y_M, ..., y_N) \geq h(x_M, ..., x_N) - \epsilon .$$

It is easily seen that this is equivalent to the conclusion of the Theorem. $\square$

We will also use the notation

$$(13.4) \qquad \Delta h_{q,a}(y,x) = \lim_{M\to-\infty,\, N\to\infty} \sum_{a+Mq}^{a+Nq-1} (h(y_i, y_{i+1}) - h(x_i, x_{i+1})) ,$$

when this limit exists (we admit the possibility of $+\infty$ or $-\infty$ as the limit). If this limit exists for every a and is independent of a, we will say that $\Delta h_q(y,x)$ exists, and write $\Delta h_q(y,x)$ for the common value.

For example, if both x and y are minimal configurations of type (p,q), then $\Delta h_q(y,x) = 0$, whereas $\Delta h(y,x)$ does not exist.

Theorem 13.2. *Let x and ξ be periodic minimal configurations of type (p, q). Let y be a configuration such that $|y_i - x_i| \to 0$ as $i \to -\infty$, and $|y_i - \xi_i| \to 0$ as $i \to +\infty$. Then $\Delta h_q(y, x) = \Delta h_q(y, \xi)$ exists. Moreover, y is minimal if and only if y minimizes this quantity over all such configurations.*

Proof. For every $\epsilon > 0$, there exists N_0 such that

$$h(y_M, ..., y_N) \geq h(x_M, ..., x_N) - \epsilon$$

if $M, N \leq N_0$ and

$$h(y_M, ..., y_N) \geq h(\xi_M, ..., \xi_N) - \epsilon$$

if $M, N \geq N_0$, just as in the proof of Theorem 13.1. This, together with $\Delta h_q(x, \xi) = 0$ implies that $\Delta h_q(y, x)$ and $\Delta h_q(y, \xi)$ exist and are equal.

Clearly, if y minimizes $\Delta h_q(y, x)$ then it is minimal. The opposite implication follows easily from (12.4) (which is valid for minimal x) and the uniform continuity of h over the region defined by (12.4). $\qquad\square$

Theorem 13.3. *Let x be a minimal configuration of rotation symbol $p/q+$ (resp. $p/q-$). Then there is no minimal periodic configuration y which satisfies $l_\alpha x < y < l_\omega x$ (resp. $l_\omega x < y < l_\alpha x$).*

Proof. For definiteness, we suppose that x has rotation symbol $p/q+$, the other case being similar. If y is a periodic minimal configuration which satisfies $l_\alpha x < y < l_\omega x$, then it is periodic of type (p, q) and $\Delta h_q(y, l_\alpha x) = \Delta h_q(y, l_\omega x) = 0$. The following inequality is analogous to (9.1) and also follows immediately from (H_3):

$$(13.5) \qquad \Delta h_q(x \vee y, y) + \Delta h_q(x \wedge y, y) \leq \Delta h_q(x, y) .$$

Furthermore, just as with (9.1), if $G(x)$ crosses $G(y)$ in the interior of an interval, then this inequality is strict.

Let i_0 be such that $x_i \leq y_i$ for $i \leq i_0$ and $x_i > y_i$ otherwise. Let

$$z_i = \begin{cases} x_i , & i \leq i_0 \\ y_0 , & i_0 < i \leq i_0 + q \\ x_{i-q} , & i > i_0 + q . \end{cases}$$

It is easily verified that

$$\Delta h_q(z, y) = \Delta h_q(x \vee y, y) + \Delta h_q(x \wedge y, y) .$$

Thus, $\Delta h_q(z, y) \leq \Delta h_q(x, y)$ and this inequality is strict if $G(x)$ crosses $G(y)$ in the interior of an interval. But, the assumption that x is minimal implies that $x = \xi$ minimizes $\Delta h_q(\xi, y)$ subject to the conditions $|\xi_i - (l_\alpha x)_i| \to 0$ as $i \to -\infty$ and $|\xi_i - (l_\omega x)_i| \to 0$ as $i \to +\infty$. This gives a contradiction in the case that $G(x)$ crosses $G(y)$ in the interior of an interval.

In the remaining case, we may obtain a contradiction by using (H_4). For, there exists z^* such that $z_i^* = z_i$, if $i \neq i_0$, and $\Delta h_q(z^*, y) < \Delta h_q(z, y) = \Delta h_q(x, y)$, since $(y_{i_0-1}, y_{i_0}, y_{i_0+1})$ is a minimal segment, and therefore (H_4) implies that the segment $(z_{i_0-1}, z_{i_0}, z_{i_0+1})$ is not minimal, as $y_{i_0-1} < z_{i_0-1}$, $y_{i_0} = z_{i_0}$, $y_{i_0+1} = z_{i_0+1}$. $\qquad\square$

By the *symbol space* S we will mean the disjoint union of the real numbers $\mathbf{R}$ together with the symbols $p/q-$ and $p/q+$, where p/q ranges over the rational numbers. By the *projection* $\pi : S \to \mathbf{R}$, we mean $\pi(\omega) = \omega$, if $\omega \in \mathbf{R}$, and $\pi(p/q\pm) = p/q$. We will also call $\pi(\omega)$ the *underlying number* of the symbol ω. We provide S with an order as follows: if $\pi(\omega_0) < \pi(\omega_1)$, then $\omega_0 < \omega_1$, and $p/q- < p/q < p/q+$. We provide S with the order topology.

By the *rotation symbol* $\tilde{\rho}(x)$ of a minimal configuration x, we mean the rotation number $\rho(x)$ when this is irrational and what we have already defined it to be when $\rho(x)$ is rational. Thus, to every h-minimal configuration x, we have associated a rotation symbol $\tilde{\rho}(x) \in S$, whose underlying number is $\rho(x)$.

For every rotation symbol ω, we have defined a set $\mathcal{M}_\omega = \mathcal{M}_\omega^h \subset \mathcal{M}^h$: when ω is an irrational number, this was defined in §12; when $\omega = p/q-$, p/q or $p/q+$, in this section. By definition, $\mathcal{M}_{p/q} = \mathcal{M}_{p/q-} \cap \mathcal{M}_{p/q+}$.

Trichotomy IV. *If ω is a rotation symbol and $x,y \in \mathcal{M}_\omega^h$, then $x < y$, $y < x$ or $x = y$.*

Proof. For the case that ω is an irrational number, this is Trichotomy II. For the case that ω is a rational number, this is an easy consequences of Aubry's Crossing Lemma. The only remaining cases are $\omega = p/q-$ or $p/q+$. We suppose that $\omega = p/q+$, the other case being similar. Suppose x, $y \in \mathcal{M}_\omega^h$. If $\tilde{\rho}(x) = \tilde{\rho}(y) = p/q$, we have already excluded crossing. If $\tilde{\rho}(x) = p/q+$, we have $l_\alpha x < x < l_\omega x$ by the Addendum to Aubry's Crossing Lemma. If furthermore $\tilde{\rho}(y) = p/q$, crossing is excluded by Theorem 13.3. On the other hand, if $\tilde{\rho}(x) = p/q+$, it follows from Theorem 13.3 that crossing implies that $l_\alpha x = l_\alpha y$ and $l_\omega x = l_\omega y$. Then crossing is excluded by the Addendum to Aubry's Crossing Lemma. $\qquad\square$

We set $A_\omega = A_\omega^h = \mathrm{pr}_0(\mathcal{M}_\omega^h)$, for any rotation symbol ω.

Theorem 13.4. *A_ω^h is a closed subset of $\mathbf{R}$, and $\mathrm{pr}_0 : \mathcal{M}_\omega^h \to A_\omega^h$ is a homeomorphism, for any $\omega \in S$.*

Proof. This follows from Trichotomy IV in the same way as Theorem 12.2 follows from Trichotomy II. $\qquad\square$

Note that Trichotomy II and III are special cases of Trichotomy IV. It is, however, necessary to consider these special cases for the proofs.

Finally, we consider the existence theory for configurations of rotation symbol $p/q\pm$. We have already obtained the existence theory for other rotation symbols: by Corollary 10.4, $A_{p/q} \neq \emptyset$, and by Theorem 11.5, $A_\omega \neq \emptyset$, if ω is an irrational number.

Theorem 13.5. *If J is a complementary interval of $A_{p/q}$ (i.e. a component of $\mathbf{R} \setminus A_{p/q}$), then $J \cap A_{p/q+} \neq \emptyset$ and $J \cap A_{p/q-} \neq \emptyset$.*

Proof. We will show that $J \cap A_{p/q+} \neq \emptyset$, the other case being similar.

We will show more: We let $x^{(1)}$, $x^{(2)}$, ... be minimal configurations such that $\rho(x^{(i)}) > p/q$ and $\rho(x^{(i)}) \to p/q$ as $i \to +\infty$. We let $E^{(i)} \subset \mathcal{M}$ denote the set of all translates of $x^{(i)}$. We let $L = \limsup E^{(i)}$, i.e.

$$L = \{x \in \mathcal{M} \mid \text{for every } i_0 \text{ and every neighborhood } N \text{ of } x \text{ in } \mathcal{M},$$
$$\text{there exists } i > i_0 \text{ such that } N \cap E^{(i)} \neq \emptyset\} \, .$$

We will show that $J \cap \mathrm{pr}_0(L) \neq \emptyset$. This will be clearly enough, since clearly $L \subset \mathcal{M}_{p/q+}$.

Suppose, to the contrary, that $J \cap \mathrm{pr}_0(L) = \emptyset$. Let $[a, b]$ be a compact subinterval of J. Then $E^{(i)} \cap [a, b] = \emptyset$ for all sufficiently large i. Since $\rho(x^{(i)}) > p/q$, we may suppose, by replacing $x^{(i)}$ by a suitable translate, if necessary, that $x_0^{(i)} < a$ and $x_q^{(i)} > b + p$. By passing to a subsequence, if necesssary, we may suppose that $x^{(i)}$ converges to $\xi \in \mathcal{M}_{p/q+}$. Then $\xi_0 \leq a$ and $\xi_q \geq b+p$. In fact, setting $J = (J^-, J^+)$, where $J^- < a < b < J^+$, we have $\xi_0 \leq J^-$ and $\xi_q \geq J^+ + p$, since $J \cap \mathrm{pr}_0(L) = \emptyset$. Now $J^- = \gamma_0$, $J^+ = \delta_0$, where γ, $\delta \in \mathcal{M}_{p/q}$ and $\xi_0 \leq J^-$, $\xi_q \geq J^+ + p$ imply that $l_\alpha \xi < \gamma < \delta < l_\omega \xi$, contrary to Theorem 13.3. $\qquad \square$

Corollary 13.6. *If $J = (J^-, J^+)$ is a complementary interval of $A_{p/q}$ in $\mathbf{R}$, with $J^- = \gamma_0$, $J^+ = \delta_0$, γ, $\delta \in \mathcal{M}_{p/q}$, then there exists $x \in \mathcal{M}_{p/q+}$ (resp. $y \in \mathcal{M}_{p/q-}$) with x_0 (resp. y_0) $\in J$ and $l_\alpha x = \gamma$, $l_\omega x = \delta$ (resp. $l_\alpha y = \delta$, $l_\omega x = \gamma$).*

Finally, we note that if x and y are as in this Corollary, then the Aubry graphs of x and y cross. Thus, we cannot extend trichotomy to the case of configurations of the same rotation number.

§14. Application to Dynamics.

In §§9-13, we have discussed basic results in the Aubry-Le Daeron, Bangert theory. In this section, we discuss how these results apply to dynamics.

We suppose that $(U, f) \in J$. We let $h : V \to \mathbf{R}$ denote the variational principle associated to f (§5). If $V = \mathbf{R}^2$ the application to dynamics is very simple. For example, $V = \mathbf{R}^2$ in the case of a Chirikov mapping and, more generally, if f maps $\mathbf{S}^1 \times \mathbf{R}$ onto itself, and f twists each end of $\mathbf{S}^1 \times \mathbf{R}$ infinitely. As we pointed out in §4, in this case the Bangert conditions $(H_0) - (H_4)$ are satisfied, and these are the only conditions that we used in §§9-13; we did not use (H_5) or (H_6) there.

We have already pointed out the connection to dynamics in §3 in our discussion of the Frenkel-Kontorova model. A minimal configuration $(..., x_i, ...)$ is *stationary* i.e. $\partial/\partial x_i(h(x_{i-1}, x_i) + h(x_i, x_{i+1})) = 0$. Setting $y_i = \partial_1 h(x_i, x_{i+1}) = \partial_2 h(x_{i-1}, x_i)$, we have $\tilde{f}(x_i, y_i) = (x_{i+1}, y_{i+1})$, where $\tilde{f}$ is the lift of f to the universal cover, since (5.2) is equivalent to $\tilde{f}(x, y) = (x', y')$. This provides a one-one correspondence between minimal configurations and orbits of $\tilde{f}$, in the case that f maps $\mathbf{S}^1 \times \mathbf{R}$ onto itself, $(\mathbf{S}^1 \times \mathbf{R}, f) \in J$, and f twists each end of $\mathbf{S}^1 \times \mathbf{R}$ infinitely.

We will use the expression *minimal orbits* to describe orbits of $\tilde{f}$ which corresponds to minimal configurations and also for orbits of f which lift to minimal orbits of $\tilde{f}$. We let $M = M_{\tilde{f}} \subset \mathbf{R}^2$ denote the union of minimal orbits of $\tilde{f}$ and $\Sigma = \Sigma_f \subset \mathbf{S}^1 \times \mathbf{R}$ the union of minimal orbits of f. If $\mathcal{O} = (..., (x_i, y_i), ...)$ is a minimal orbit of $\tilde{f}$, we will define the *rotation symbol* $\tilde{\rho}(\mathcal{O})$ and *rotation number* $\rho(\mathcal{O})$ to be $\tilde{\rho}(x)$ and $\rho(x)$, resp., where $x = (..., x_i, ...)$ is the corresponding minimal configuration. We will also use these terms for orbits of f. There is a slight abuse of terminology here, because changing the lift $\tilde{f}$ of f (by adding $(n, 0)$, where n is an integer) changes $\tilde{\rho}(\mathcal{O})$ and $\rho(\mathcal{O})$ (by adding n), but this should cause no confusion, since (to the authors's knowledge) all results depend only on the rotation symbol (or rotation number) mod.1.

For $\omega \in \mathcal{S}$, we let $M_\omega = M_{\omega, \tilde{f}}$ (resp. $\Sigma_\omega = \Sigma_{\omega, \tilde{f}}$) be the subset of M (resp. Σ) consisting of orbits of rotation symbol ω. From Theorem 13.4, it follows that M_ω (resp. Σ_ω is a closed subset of $\mathbf{R}^2$ (resp. $\mathbf{S}^1 \times \mathbf{R}$), and the projection π_1 on the first factor induces a homeomorphism $\pi_1 : M_\omega \cong A_\omega$ (resp. $\pi_1 : \Sigma_\omega \cong A_\omega/\mathbf{Z}$), which satisfies the following property:

Theorem 14.1. *If the variational principle h is C^2 (i.e. f is C^1) and satisfies the Bangert conditions $(H_1) - (H_4)$, then the homeomorphism $\pi_1 : \Sigma_\omega \to A_\omega/\mathbf{Z}$ has Lipschitz inverse, whose Lipschitz constant only depends on h and its second derivatives.*

Proof. Let $\mathrm{pr}_0 : \mathcal{M}_\omega^h \to A_\omega^h$ be the homeomorphism given in Theorem 13.4 and let S be the shift map of $\mathcal{M}_\omega$ which has been introduced in §12. Then the map $\Phi : A_\omega \to A_\omega$, $\Phi = \mathrm{pr}_0^{-1} \circ S \circ \mathrm{pr}_0$ is bi-Lipschitz. In fact, since $|x_1 - x_0|$ and $|x_0 - x_{-1}|$ are uniformly bounded for all $x \in \mathcal{M}_\omega^h$ and by (H_1) it is sufficient to consider the case when $0 \leq x \leq \tilde{x} < 2$, we may assume that $\Phi^{-1}(x)$, x, $\Phi(x)$ and $\Phi^{-1}(\tilde{x})$, $\tilde{x}$, $\Phi(\tilde{x})$ are contained in a fixed compact interval I. By the Bangert properties and the differentiability of h, there exist $\delta > 0$ and $L > 0$ such that $\partial_{12} h \leq -\delta < 0$ on $I \times I$

and $\partial_1 h$, $\partial_2 h$ are Lipschitz on $I \times I$ with constant L. Then the following estimate holds:

$$0 \le \delta\left(\Phi^{-1}(\tilde{x}) - \Phi^{-1}(x)\right) + \left(\Phi(\tilde{x}) - \Phi(x)\right) \le \partial_2 h(\Phi^{-1}(x), \tilde{x}) - \partial_2 h(\Phi^{-1}(\tilde{x}), \tilde{x}) +$$
$$+ \partial_1 h(x, \Phi(x)) - \partial_1 h(x, \Phi(\tilde{x})) = \partial_2 h(\Phi^{-1}(x), \tilde{x}) - \partial_2 h(\Phi^{-1}(x), x) +$$
$$+ \partial_1 h(\tilde{x}, \Phi(\tilde{x})) - \partial_1 h(x, \Phi(\tilde{x})) \le 2L(\tilde{x} - x) ,$$

where the first inequality is a consequence of the Aubry's Crossing Lemma, the second and the last one follow from the mean value theorem and the equality in between holds since $(\Phi^{-1}(x), x, \Phi(x))$ and $(\Phi^{-1}(\tilde{x}), \tilde{x}, \Phi(\tilde{x}))$ are stationary segments with respect to h. This proves our claim. Finally, $\pi_1^{-1} : A_\omega \to M_\omega$ can be written as

$$\pi_1^{-1}(x) = (x, -\partial_1 h(x, \Phi(x))) ,$$

which, in view of the previous argument, implies the statement of the theorem. $\qquad\square$

We also get commutative diagrams of homeomorphisms:

$$\begin{array}{ccc} M_{\omega,\tilde{f}} & \overset{\tilde{f}}{\to} & M_{\omega,\tilde{f}} \\ \downarrow & & \downarrow \\ \mathcal{M}_\omega^h & \overset{S}{\to} & \mathcal{M}_\omega^h \end{array} \qquad\qquad \begin{array}{ccc} \Sigma_{\omega,f} & \overset{f}{\to} & \Sigma_{\omega,f} \\ \downarrow & & \downarrow \\ \mathcal{M}_\omega^h/T & \overset{\overline{S}}{\to} & \mathcal{M}_\omega^h/T \end{array}$$

For example, the commutativity of the second diagram, together with the fact that the vertical arrows are homeomorphisms, says that f is topologically conjugate to $\overline{S}$. Thus Theorem 12.3 may be restated as:

Theorem 14.2. *If ω is an irrational number, then $f : \Sigma_{\omega,f} \to \Sigma_{\omega,f}$ has a unique minimal set (in the sense of topological dynamics). Every orbit of $f|\Sigma_{\omega,f}$ tends to the minimal set under forward and backward iteration. Moreover, $f : \Sigma_{\omega,f} \to \Sigma_{\omega,f}$ is semi-conjugate to a rotation.*

Note that in §12, we assumed the standing hypothesis that ω is an irrational number. The hypothesis that ω is an irrational number is essential: if $\omega = p/q-$, p/q, $p/q+$, the minimal sets in $\Sigma_{\omega,f}$ are the periodic orbits (whose union is $\Sigma_{p/q,f}$), and every orbit tends to $\Sigma_{p/q,f}$ under forward and backward iteration.

The embedding $A_{\omega,\tilde{f}} \subset \mathbf{R}$ induces a total order on $A_{\omega,\tilde{f}}$, which may be used to define a total order on $M_{\omega,\tilde{f}}$ *via* the homeomorphism $\pi_1 : M_{\omega,\tilde{f}} \to A_{\omega,\tilde{f}}$. Trichotomy IV has the following further consequence: for any rotation symbol ω the mapping $\tilde{f} : M_{\omega,\tilde{f}} \to M_{\omega,\tilde{f}}$ preserves this order.

Likewise, we may define a cyclic order on $\Sigma_{\omega,f}$ *via* the homeomorphism $\pi_1 : \Sigma_{\omega,f} \to A_{\omega,f}/\mathbf{Z} \subset \mathbf{S}^1 = \mathbf{R}/\mathbf{Z}$. We have the further consequence that $f : \Sigma_{\omega,f} \to \Sigma_{\omega,f}$ preserves the cyclic order.

Now we may explain the relation between the Aubry-Le Daeron, Bangert method, which we have been describing in the last several sections and the first author's method, which we described in §6. We recall that to a minimal configuration x of irrational rotation symbol we associated (following Aubry) in §12 a *hull function* ϕ_x.

Theorem 14.3. *If ω is an irrational number and x is a minimal configuration of rotation number ω, then the hull function ϕ_x minimizes $F_\omega(\phi)$ over the set of measurable functions ϕ which satisfy the periodicity condition $\phi(t+1) = \phi(t)+1$ and the condition that $\phi(t) - t$ should be bounded. Moreover, the minimizer is unique up to translation, i.e. if ϕ minimizes F_ω, then $\phi = \phi\, T_a$ almost everywhere, for some $a \in \mathbf{R}$, where T_a is the translation $T_a(t) = t + a$.*

In fact, this holds for any variational principle h which satisfies $(H_0) - (H_4)$; it is not necessary to assume that h is the variational principle of a mapping. We will not give a proof. For twist mappings, this result was proven [Ma6, Prop. 11.1]; the same proof works for the formulation we have given here.

It follows that the set Σ_ω^* defined in §6 is the unique minimal set (in the sense of topological dynamics) of $f|\Sigma_\omega$, in the case that ω is irrational.

§15. Birkhoff Invariant Curve Theorem.

In the next sections, we will be concerned with the properties of a special but very important class of invariant sets for exact area-preserving monotone twist maps of the annulus $S^1 \times R$: *rotational* invariant curves, i.e. invariant curves which separate the top from the bottom of the annulus. We will see that the action minimizing sets Σ_ω introduced and studied in the previous sections, are a generalization of rotational invariant curves since, as it will be shown in §17, every rotational invariant curve is an action minimizing set. Furthermore, if Γ is a rotational invariant curve which is invariant for a diffeomorphism $f \in J^1$, then clearly $f|\Gamma$ is topologically conjugate to an orientation preserving homeomorphism of the circle S^1 and thus it has a well defined *rotation number* $\rho(\Gamma) = \omega \in R$. Then $\Gamma = \Sigma_\omega$ in case ω is irrational and $\Gamma = \Sigma_{p/q\pm}$, if $\omega = p/q$. The basic result on rotational invariant curves is the following theorem due to G.D. Birkhoff [Bi1], [Bi3], which establishes in the case of curves the analogous of Theorem 14.1:

Theorem 15.1. *Let $f \in J^1$ be an area-preserving monotone twist diffeomorphism of the annulus $S^1 \times R$. Then every rotational invariant curve Γ is the graph of a Lipschitz function $u : S^1 \to R$, i.e. it has the following form $\Gamma = \{(\theta, u(\theta)) \,|\, \theta \in S^1\}$. Furthermore, the Lipschitz constant of u, $Lip(u)$, can be a priori estimated from the properties of the diffeomorphism f.*

This theorem is a particular case of a more general result:

Theorem 15.1'. *Let $f : S^1 \times R \to S^1 \times R$ be a C^1 diffeomorphism. Suppose that f preserves the area, maps each end of $S^1 \times R$ to itself, preserves orientation, and deviates vertical lines $\{(\theta, y) \in S^1 \times R \,|\, \theta = \text{ constant }\}$ in $S^1 \times R$ either to left or to the right. Let U be an open subset of $S^1 \times R$ such that $fU = U$, U is homeomorphic to $S^1 \times R$ and $S^1 \times (-\infty, a] \subset U \subset S^1 \times (-\infty, b)$, for some $a < b$, a, $b \in R$. Then the frontier of U in $S^1 \times R$ is the graph of a Lipschitz function $u : S^1 \to R$, i.e. $\overline{U} \setminus U = \{(\theta, u(\theta)) \,|\, \theta \in S^1\}$. Furthermore, the Lipschitz constant of u, $Lip(u)$, can be a priori estimated from the properties of the diffeomorphism f.*

Example 15.2. Suppose that U satisfies all hypotheses of Theorem 15.1' but the one of being homeomorphic to $S^1 \times R$. Then Theorem 15.1' can be applied to the open set $\tilde{U}$ obtained from U by "filling the holes", i.e. $\tilde{U}$ is the complement in $S^1 \times R$ of the unbounded connected component of the complement of U.

Example 15.3. Let $\Gamma \subset S^1 \times R$ be a rotational invariant curve for a diffeomorphism $f \in J^1$. Then f satisfies the hypotheses of the theorem and one of the components of $S^1 \times R \setminus \Gamma$ satisfies the condition imposed on U (as a consequence of Schoenflies theorem). Thus Theorem 15.1 follows easily from Theorem 15.1'.

Proof. The reader can consult [Bi1, §44], [Bi3, §3], [He2, Chap. I], [Ma4, §2] for further details.

A point $x \in U$ will be said to be *positively* (resp. *negatively*) *accessible* if there is an embedded curve $\gamma : (-\infty, a) \to U$ satisfying $\gamma(t)_2 \to -\infty$, as $t \to -\infty$, and having positive (resp. negative) deviation from the vertical, such that $\gamma(a) = x$. We let W_+ (resp. W_-) denote the set of positively (resp. negatively) accessible points in U.

Assume f deviates vertical lines to the right (the case when f deviates vertical lines to the left can be reduced to the this case by replacing f with f^{-1}. In the case f deviates vertical line to the right, clearly

$$(15.1) \qquad f(\overline{W}_- \cap U) \subset W_- \quad \text{and} \quad f^{-1}(\overline{W}_+ \cap U) \subset W_+ \, .$$

Since f is area-preserving and there exists $a < b$ such that

$$\mathbf{S}^1 \times (-\infty, a] \subset W_- \cap W_+ \cap U \subset W_- \cup W_+ \cup U \subset \mathbf{S}^1 \times (\infty, b] \, ,$$

then (15.1) implies $W_- = W_+ = U$. In fact, $U \setminus W_-$ and $U \setminus W_+$ are closed sets in U, have finite area and, by (15.1),

$$U \setminus W_- \subset U \setminus f(\overline{W}_- \cap U) = f(U \setminus \overline{W}_-) \, ,$$

which contradicts the area-preserving property, unless $W_- = U$. Similarly, $W_+ = U$.

As a consequence, for each $\theta \in \mathbf{S}^1$, there exists a_θ such that

$$(15.2) \qquad U \cap (\{\theta\} \times \mathbf{R}) = \{\theta\} \times (-\infty, a_\theta) \, .$$

The argument goes as follows: if (15.2) is false, there exist two points $A, B \in \partial U$ such that the segment $[A, B]$ is vertical and does not intersect ∂U. Therefore $[A, B]$ divides U into two connected components: U_1, homeomorphic to $\mathbf{S}^1 \times \mathbf{R}$, and U_2, homeomorphic to $\mathbf{R}^2$. This follows from the Jordan-Schoenflies theorem. If U_2 lies on the left of the vertical interval $[A, B]$, then U_2 is not contained in W_-, similarly, it is not contained in W_+, if it lies on the right of $[A, B]$. In either case we obtain a contradiction with the fact that $W_- = W_+ = U$.

The conclusion (15.2) already shows that the frontier of U, ∂U, is the graph of a function. Furthermore, if $u : \mathbf{R} \to \mathbf{R}$ is a C^1 function, we let $\phi_u : \mathbf{S}^1 \times \mathbf{R} \to \mathbf{S}^1 \times \mathbf{R}$ be the C^1 diffeomorphism

$$\phi_u(\theta, y) = (\theta + u(y), y) \, .$$

Let $f_u = \phi_u \circ f \circ \phi_u^{-1}$. Then $\phi_u(U)$ is invariant under f_u and, if u is sufficiently small in the C^1 Whitney topology, f_u deviates vertical lines the same side as f. Therefore $\phi_u(U)$ satisfies the conditions required to follow the above arguments to the conclusion that, for each $\theta \in \mathbf{S}^1$, there exists $a_\theta(u) \in \mathbf{R}$ such that

$$\phi_u(U) \cap (\{\theta\} \times \mathbf{R} = \{\theta\} \times (-\infty, a_\theta(u)) \, .$$

Since this is true for any u sufficienlty small in the C^1 topology, it follows that the frontier of U is the graph of a Lipschitz function, which is the content of Birkhoff theorem. The last part of the argument also shows that the Lipschitz constant can be estimated *a priori* from the properties of the diffeomorphism f and it does not depend on the invariant set U. In fact it will depend on the size of the neighborhood $\mathcal{U}$ of f in the C^1 topology such that any $g \in \mathcal{U}$ still deviates vertical lines the same side as f does. $\qquad \square$

§16. A Survey of K.A.M. Theory

The Birkhoff invariant curve theorem can be thought as a regularity result for invariant curves. It establishes that every *rotational* invariant curve for a monotone-twist exact area-preserving diffeomorphism of the annulus must be the graph over S^1 of a Lipschitz function. Thus, it is natural to ask what is the minimal regularity of rotational invariant curves for a smooth or analytic diffeomorphism. In [S-Z], D.Salamon and E.Zehnder proved that, if the rotation number of the invariant curve is strongly irrational and the rotational invariant curve is sufficiently differentiable, then it is as smooth as the diffeomorphism f, i.e. it is C^∞ if f is C^∞, it is analytic if f is analytic. This result, whose flavour is certainly close to hypoellipticity for P.D.E., has been obtained by a K.A.M. iteration method in configuration space, i.e. for the Euler-Lagrange equation associated to a Lagrangian system.

We recall that, by a result due to J. Moser [Mo2], any exact area-preserving monotone twist map f of the annulus can be interpolated by a time dependent (non-autonomous) periodic Hamiltonian flow, induced by a Hamiltonian $H : T^*S^1 \times \mathbf{R} \times \mathbf{R} \to \mathbf{R}$, satisfying the Legendre condition

$$H_{yy}(\theta, y, t) > 0 \ ,$$

i.e. f coincides with the time-1-map of the Hamiltonian flow associated to H. Since H satisfies the Legendre condition, f can also be interpolated by the time-1-map associated to the Lagrangian flow which can be obtained from the previous Hamiltonian flow by the usual Legendre transformation.

The K.A.M. theory has been originally found by Kolmogorov and Arnold in the analytic case, by Moser in the smooth case as a perturbative method to establish the existence of invariant tori for Hamiltonian systems (close to be integrable). The differentiability requirement have been weakened from the famous 333 derivatives of the original paper by Moser to the optimal condition requiring 3 derivatives by the efforts various mathematicians as Pöschel, Rüssmann, Herman. As the same time, the estimates on the size of the perturbation allowed have been improved. We refer to the second volume of Herman book [He] (for the special case of invariant curves), to the paper by Salamon and Zehnder [S-Z] and references therein. On the other hand, methods to prove converse results in the case of rotational invariant curves for monotone twist mappings have been developed by Herman [He], who in particular proved the optimality of the condition requiring the smallness of the perturbation in the C^3 topology, and later extended by the first author [Ma4], [Ma7], [Ma8] and [Ma9] and by the second author [F]. We will give an exposition of these results in §17 and §18.

We will now state the K.A.M. theorem for invariant curves in the version of [S-Z]. A real irrational number ω is said to satisfy a *Diophantine condition* if there exist constants $C > 0$ and $\tau \geq 1$ such that

(16.1) $$|q\omega - p| \geq C/q^\tau \quad \text{for any} \ p \in \mathbf{Z} \ \text{and} \ q > 0 \ .$$

This condition express the property of ω being badly approximated by rationals. If ω does not satisfy this condition, it is said to be a *Liouville* number. We recall that the sets of Diophantine and Liouville numbers are dense in $\mathbf{R}$, but the latter is of Hausdorff dimension and Lebesgue measure zero, as it is not difficult to show.

Theorem 16.1. *[S-Z] Let $f_0 \in J^1$ be an exact area-preserving monotone twist diffeomorphism of the annulus $\mathbf{S}^1 \times \mathbf{R}$. Let γ_0 be a rotational invariant curve for f_0 of rotation number $\omega \in \mathbf{R} \setminus \mathbf{Q}$, satisfying the Diophantine condition (16.1). Suppose that f_0 is of class C^∞ and $f_0|\gamma_0$ is C^{l+1}-conjugate to the rotation $R_\omega : \mathbf{S}^1 \to \mathbf{S}^1$, $l > 2\tau + 2$ and $l - 2\tau - 1$, $l - \tau - 1$ are not integers. Then $f_0|\gamma_0$ is C^∞-conjugate to the rotation R_ω. In addition, if f_0 is analytic, the conjugacy is analytic. Furthermore, there exists a neighborhood $\mathcal{U} = \mathcal{U}(C, \tau, f_0, \gamma_0)$ of f_0 in the C^l topology such that every exact area-preserving monotone twist C^∞ diffeomorphism of the annulus $f \in \mathcal{U}$ admits a rotational invariant curve γ of rotation number ω.*

In the *analytic* case, H.Rüssmann has shown in a series of papers, concluded in [Rs], that the Diophantine condition (16.1) on the rotation number $\omega \in \mathbf{R} \setminus \mathbf{Q}$ can be replaced in Theorem 16.1 by a weaker condition. Let $(p_n/q_n)_{n \in \mathbf{N}}$ the sequence of convergents of the continued fraction expansion of ω. The condition (16.1) can be expressed in terms of the continued fraction expansion as

$$(16.1') \qquad \mathrm{Log}\, q_{n+1} \leq C \mathrm{Log}\, q_n \quad, \quad n \in \mathbf{N} .$$

Rüssmann proves that the condition

$$(16.2) \qquad \sum_{n \in \mathbf{N}} \frac{\mathrm{Log}\, q_{n+1}}{q_n} < \infty ,$$

previously introduced by A.D. Brjuno in connection with problems related to classical perturbation theory in Hamiltonian mechanics and known as the *Brjuno condition*, it is sufficient for the stability result of invariant curves contained in Theorem 16.1: if γ_0 is analytically conjugate to a rotation and its rotation number $\omega \in \mathbf{R} \setminus \mathbf{Q}$ satisfies the Brjuno condition (16.2), then there exists a neighborhood $\mathcal{U}$ of the map f_0 in the analytic topology, such that any exact area-preserving monotone twist diffeomorphism $f \in \mathcal{U}$ admits a rotational invariant curve γ of rotation number ω. Clearly the Diophantine condition (16.1) implies the Brjuno condition (16.2), thus Rüssmann theorem is in fact stronger than Theorem 16.1 in the analytic case.

We recall the basic facts known about the relations between the smoothness of the invariant curve γ as a curve in $\mathbf{S}^1 \times \mathbf{R}$ and the smoothness of the conjugacy of the diffeomorphism $f|\gamma$ to the corresponding rigid rotation of the circle. It is a classical counterexample by Arnold that there exist analytic diffeomorphisms of the circle, with irrational rotation number ω, whose conjugacy to a rigid rotation (which exists and is a homeomorphism by a classic theorem due to Denjoy [De]) is not absolutely continuous. Analogous examples have been constructed in the smooth case. Since every diffeomorphism of the circle can be embedded as rotational invariant curve of an exact area-preserving monotone twist diffeomorphism of the annulus with the same degree of smoothness, the mentioned examples give examples of analytic (resp. smooth) monotone twist maps f having an analytic (resp. smooth) rotational invariant curve γ such that $f|\gamma$ is not absolutely continuously conjugate to a rigid rotation (although topologically it is, by Denjoy theorem). However, in these examples the rotation number ω, although irrational, is very well approximated by rationals.

On the other hand, in the case of rotation numbers satisfying a Diophantine condition, Herman's theorem holds. We state the improved version due to J.C. Yoccoz [Yo]:

Theorem 16.2. Let $\phi : S^1 \to S^1$ be a diffeomorphism of the circle of class C^k, $k \in \mathbf{N}$ and $k \geq 3$. Suppose that the rotation number $\omega \in \mathbf{R} \setminus \mathbf{Q}$ satisfies the Diophantine condition (16.1). Then, if $k > 2\tau - 1$, ϕ is $C^{k-\tau-\epsilon}$-conjugate to the rigid rotation R_ω, for any $\epsilon > 0$. In addition, if ϕ is C^∞ the conjugacy is C^∞, if ϕ is real analytic the conjugacy is also analytic.

Theorem 16.2 implies that, if γ_0 is a rotational invariant curve of rotation number $\omega \in \mathbf{R} \setminus \mathbf{Q}$, satisfying the Diophantine condition (16.1), then, if γ_0 is sufficiently smooth *as a curve* in $S^1 \times \mathbf{R}$ then Theorem 16.1 applies.

In the analytic case, the Diophantine condition in Theorem 16.2 can be replaced by weaker conditions. The picture of the situation, due to J.-C. Yoccoz, is described in the survey paper by R.Perez-Marco [P-M]. We just mention that there exists a condition $\mathcal{H}$ on the rotation number ω, weaker than the Diophantine condition (16.1) but stronger than the Brjuno condition (16.2), such that, if ω satisfies $\mathcal{H}$, then the analytic diffeomorphism of the circle ϕ is analytically conjugate to its linear part R_ω. On the other hand, if ω does not satisfies the Brjuno condition, then there are examples of analytic diffeomorphisms ϕ which are not analytically conjugate to R_ω. However, there is a significant gap between the condition $\mathcal{H}$ and the Brjuno condition. Thus, the situation is, at the level of optimal conditions, more delicate than in the smooth case.

One of the most interesting applications of K.A.M. theorem for invariant curves of exact area-preserving monotone twist diffeomorphisms of the annulus is the proof of the *stability* of elliptic periodic points of area-preserving diffeomorphisms of surfaces. The problem can be reduced to the case of a fixed point by considering the appropriate iteration of the map.

Corollary 16.3. Suppose that P is an elliptic fixed point of a C^∞ area-preserving mapping f of an open subset of the plane into the plane. Suppose furthermore that the Birkhoff invariants (see §4) of the map f at P are not all equal to zero. Then P is Lyapunoff stable.

Proof. According to the Birkhoff normal form theorem, stated in §4, there exists an area-preserving change of coordinates in a neighborhood of P such that the mapping f takes the form

$$f(\zeta) = \zeta \exp 2\pi i(\beta_0 + \beta_1 \rho^2 + ... + \beta_N \rho^{2N}) + O(\rho^k)$$

and $k = 2N+2$ or $2N+3$. Here $\zeta = \xi + i\eta$ denotes the complex coordinate associated to the real coordinates ξ, η, and $\rho = (\xi^2 + \eta^2)^{1/2}$. The real numbers $\beta_1, ..., \beta_N, ...$ are called the Birkhoff invariants of f at P. If an eigenvalue λ of $df(P)$ is not a root of unity, we may take $N = \infty$. If λ is a primitive q^{th} root of unity, then we may take $k = q$ and $N = [q - 2/2]$. Thus f is the sum of the normal form

$$N(\zeta) = \zeta \, \exp 2\pi i(\beta_0 + \beta_1 \rho^2 + \ldots + \beta_N \rho^{2N})$$

and a reminder term which is no bigger than $O(\rho^k)$, where $k = \infty$ if an eigenvalue λ of $df(P)$ is not a root of unity and $k = q$ if λ is a primitive q^{th} root of unity.

If at least one of the Birkhoff invariants is not zero, then the normal form $N(\zeta)$ and f are exact area-preserving monotone twist mappings in a sufficiently small punctured neighborhood of P, as explained in §4. Furthermore, this neighborhood is foliated by invariant curves of the normal form $N(\zeta)$, since $\beta_0, \beta_1, \ldots$ are real numbers, and the set of rotation numbers of invariant curves of the normal form is an open interval $I \subset \mathbf{R}$, since at least one of the Birkhoff invariants is non-zero. Therefore, by K.A.M. theorem (Theorem 16.1), if an eigenvalue λ of $df(P)$ is not a root of unity or at least it is not a qth root of unity for small q, there exists a sufficiently small punctured neighborhood of P, which contains a positive measure set of invariant curves of the map f. In fact, f can be considered, in a small neighborhood of P, as a small perturbation of its normal form and by K.A.M. theorem invariant curves having Diophantine rotation numbers (i.e. satisfying condition (16.1)) persist for small perturbations. Since in dimension 2 invariant curves are separating, the existence of invariant curves surrounding P implies the stability of P. $\qquad\square$

We conclude the section by stating the main application, due to Lazutkin [La], of K.A.M. theory to plane convex billiards. Let R be an open convex bounded region in the plane whose boundary is of class C^2. The billiard ball problem in R has been briefly described in §3, where it is explained the basic fact, already known to Birkhoff, that its dynamics largely reduces to the dynamics of an associated exact area-preserving monotone twist mapping. A *caustic* for the billiards ball problem in R is a closed curve $\mathcal{C}$ in R such that any trajectory which starts being tangent to $\mathcal{C}$ stays tangent to $\mathcal{C}$ after bouncing onto ∂R, forever in the past and in the future. A caustic corresponds to a rotational invariant curve for the exact area-preserving monotone twist mapping associated to the billiard ball problem.

Theorem 16.4. *[La] If R is strictly convex (i.e. the curvature of ∂R never vanishes) and ∂R is sufficiently differentiable, then there exist caustics for the billiard ball problem in R (arbitrarily close to ∂R).*

In the next section we will give a converse, due to the first author [Ma2], to Theorem 16.4. The argument will be based on the Birkhoff invariant curve theorem described in §15.

§17. Birkhoff Invariant Curve Theorem. Applications

In this section we discuss converse K.A.M. results, due to the first author [Ma2]-[Ma4], which are based on the Birkhoff invariant curve theorem described in §15. Then we prove the basic variational property of rotational invariant curves, i.e. any orbit on a rotational invariant curve for a C^1 exact area-preserving monotone twist diffeomorphism of the annulus is a *minimal orbit* in the sense established at the beginning of §14. This property has many of important consequences, which will be discussed in sections §18 and §19.

Glancing Billiards.

We present a converse [Ma2] to the Lazutkin theorem of the last section (Theorem 16.4). Let R be a convex bounded plane open region, whose boundary ∂R is C^2. In R we consider the billiard ball problem, already described in §3. We will say that a trajectory is ϵ-glancing if for at least one bounce the angle of reflection (with either the positive or negative tangent of ∂R at the point of reflection) is $< \epsilon$. If $\epsilon < \pi/2$, we can distinguish between a positively ϵ-glancing trajectory and a negatively ϵ-glancing trajectory according to whether the direction of reflection is close to the positive resp. the negative tangent to ∂R.

Theorem 17.1. *[Ma2] If the curvature of ∂R is zero at some point, then the billiard ball problem in R has no caustics. As a consequence, for any $\epsilon > 0$, there exist trajectories which are both positively and negatively ϵ-glancing.*

Proof. The proof depends on the formulation of the billiard ball problem in terms of area-preserving diffeomorphism of the annulus. A trajectory is positively and negatively ϵ-glancing if and only if the corresponding orbit for the associated exact area-preserving monotone twist mapping of the annulus $f : \partial R \times (0, \pi) \to \partial R \times (0, \pi)$ visits ϵ-neigbourhoods of both boundaries of the annulus, $\partial R \times (0, \epsilon)$ and $\partial R \times (\pi - \epsilon, 0)$. It is a classical consequence of Birkhoff invariant curve theorem, already obtained by Birkhoff himself, that the following holds:

Lemma 17.2. *Let $f : \mathbf{S}^1 \times (0, \pi) \to \mathbf{S}^1 \times (0, \pi)$ be an exact area-preserving monotone twist map. Assume f has no rotational invariant curves. Then, for any $\epsilon > 0$, if $V_- = \mathbf{S}^1 \times [0, \epsilon)$ and $V_+ = \mathbf{S}^1 \times (\pi - \epsilon, \pi]$, there exists an orbit of the map f connecting V_- and V_+, i.e. there exist $P \in \mathbf{S}^1 \times (0, \pi)$ and integer n_-, n_+ such that $f^{n_-}(P) \in V_-$ and $f^{n_+}(P) \in V_+$.*

Proof. Suppose there exists $\epsilon > 0$ for which the above statement does not hold. Chose such ϵ and consider the correspondin sets V_- and V_+. Let

$$V = \bigcup_{n \in \mathbf{Z}} f^n(\mathrm{int}\, V_-) \cup (\mathbf{S}^1 \times \{0\}) .$$

Clearly we would have

$$V \cap V_+ = \emptyset .$$

Let B be the connected component of $\mathbf{S}^1 \times [0, \pi] \setminus V$ which contains $\mathbf{S}^1 \times \{\pi\}$. Let $U = \mathbf{S}^1 \times [0, \pi] \setminus B$. Then $f(U) = U$, $V_- \subset U$, $U \cap V_+ = \emptyset$ and $\mathbf{S}^1 \times \{0\}$ is a deformation retract of U. Then, by Birkhoff invariant curve theorem (Thorem 15.1'), ∂U is the graph over $\mathbf{S}^1$ of a Lipschitz function. Therefore there exists a rotational invariant curve for f, contradicting the hypothesis. $\square$

We continue the proof of Theorem 17.1 as follows. We suppose that the curvature of ∂R is zero at some point and that there exists a rotational invariant curve for f (a caustic for the billiard ball problem). This will lead to a contradiction, thereby showing that no caustic can exists if the curvature vanishes at some point. Lemma 17.2 will complete the proof. Let $P \in \partial R$ be a point where the curvature vanishes. Let P_t a one-parameter family of points $P_t \in \partial R$ in a neighborhood of P and let $P_t^n = g^n(P_t)$, $n \in \mathbf{Z}$, where $g : \partial R \to \partial R$ describes the restriction of f to the invariant curve corresponding to the caustic. We notice that, since f is orientation preserving and also preserves the two ends of $\partial R \times [0, \pi]$, g is a orientation preserving homeomorphism of the circle. Let $h : \partial R \times \partial R \to \mathbf{R}$ be the variational principle associated to f. We have seen in §3 that h exists and has the following form:

$$h(P,Q) = -\|P - Q\| \ , \text{ for any } P, Q \in \partial R \ .$$

It is a consequence of the generating equations (3.2) and of the existence of a caustic that

(17.1) $$h_1(P_t, P_t^1) + h_2(P_t^{-1}, P_t) \equiv 0 \ ,$$

and, by differentiating with respect to t,
(17.2)
$$h_{12}(P_t, P_t^1) \frac{dP_t^1}{dt} + h_{12}(P_t^{-1}, P_t) \frac{dP_t^{-1}}{dt} = -(h_{11}(P_t, P_t^1) + h_{22}(P_t^{-1}, P_t)) \frac{dP_t}{dt} \ .$$

Since P_t^1 and P_t^{-1} represent points on the invariant curve for f associated to the caustic, by Birkhoff theorem they are Lipschitz function of t, hence their derivatives with respect to t exist almost everywhere and can be a priori bounded from above and below in terms of the mapping f. Furthermore, since g is orientation preserving $\frac{dP_t^1}{dt}$ and $\frac{dP_t^{-1}}{dt}$ have the same sign as $\frac{dP_t}{dt}$. Since $h_{12} < 0$, by the definition of a monotone twist mapping, we conclude that

(17.3) $$h_{11}(P_t, P_t^1) + h_{22}(P_t^{-1}, P_t) > 0 \ ,$$

for all $t \in \mathbf{S}^1$, as a consequence of the existence of a caustic. On the other hand, in the case of a billiard ball problem it is not difficult to show that

(17.4) $$h_{11}(P, P") + h_{22}(P', P) < 0$$

for any P', $P"$, if P is a point of ∂R where the curvature vanishes. This can be understood by considering the fact that the path of a billiard ball bouncing on a rectilinear scatterer *strictly* minimizes the euclidean length among all possible paths. Since h was defined as the negative of the euclidean length, (17.4) follows, thus contradicting (17.3) and concluding the proof of the theorem. $\square$

Non-existence of invariant curves in the Chirikov standard mapping.

The Chirikov mapping was introduced in §3 as the exact area-preserving monotone twist map f_k of the infinite cylinder $\mathbf{S}^1 \times \mathbf{R}$,

$$(17.5) \qquad f_k(\theta, y) = (\theta + y + \frac{k}{2\pi} \sin 2\pi\, \theta, y + \frac{k}{2\pi} \sin 2\pi\, \theta) \;, \quad k \in \mathbf{R} \;,$$

whose variational principle is $h_k : \mathbf{R}^2 \to \mathbf{R}$,

$$(17.5') \qquad\qquad h_k(x, x') = \frac{1}{2}(x - x')^2 - \frac{k}{4\pi^2} \cos 2\pi x \;.$$

It is in fact a one-parameter family of mappings, which is sometimes called the standard family, where k plays the role of a perturbative (stochasticity) parameter. When $k = 0$ we obtain the completely integrable mapping, which is the exact area-preserving monotone twist map f_0 of the infinite cylinder $\mathbf{S}^1 \times \mathbf{R}$ into itself, given by

$$(17.6) \qquad\qquad f_0(\theta, y) = (\theta + y, y) \;,$$

whose variational principle is $h_0 = \frac{1}{2}(x - x')^2$. This map is characterized by the property of having a rotational invariant curve for any rotation number, i.e. the cylinder $\mathbf{S}^1 \times \mathbf{R}$ is completely foliated by rotational invariant curves of the map f_0. It is a consequence of the K.A.M theorem, exposed in §16, that, when $|k|$ is sufficiently small, a large measure set of invariant curves, namely those whose rotation number satisfies the Diophantine condition (16.1), persist. On the other hand, numerical results due to Greene [Gr] show that there are rotational invariant curves for $|k| \le k_0$ and there are none for $|k| > k_0$, where the critical threshold is estimated as $k_0 \approx 0.97$. Here we will expose a simple rigorous result, due to [Ma4], which establishes that there are no rotational invariant curves for $|k| > 4/3$.

Theorem 17.3. *If $|k| > 4/3$, then there are no rotational invariant curves in $\mathbf{S}^1 \times \mathbf{R}$ which are invariant under f_k.*

Proof. The argument is similar to the one given in the proof of Theorem 17.1. Let f be an exact area-preserving monotone twist map of $\mathbf{S}^1 \times \mathbf{R}$. Suppose f has a rotational invariant curve. As a consequence of Birkhoff invariant curve theorem, the invariant curve will be the graph of a Lipschitz function $\phi : \mathbf{S}^1 \to \mathbf{R}$. Clearly, there exists an orientation preserving homeomorphism $g : \mathbf{S}^1 \to \mathbf{S}^1$ such that

$$(17.7) \qquad\qquad f(\theta, \phi(\theta)) = (g(\theta), \phi(g(\theta))) \;.$$

This follows from the fact that the curve given by the graph of the function ϕ is invariant under f. Suppose the variational principle $h : \mathbf{R}^2 \to \mathbf{R}$ of the lift of f to the universal cover $\mathbf{R}^2$ is of the following form:

$$h(x, x') = \frac{1}{2}(x - x')^2 + u(x) \;.$$

Consequently, in view of the generating equations and (17.7),

$$(17.9) \qquad \frac{dg^{-1}}{dx} + \frac{dg}{dx} - u''(x) - 2 = -\frac{d}{dx}\left(h_1(x, g(x)) + h_2(g^{-1}(x), x)\right) \equiv 0 \;.$$

Here the derivatives dg/dx and dg^{-1}/dx exist almost everywhere and are bounded, as a consequence of Birkhoff invariant curve theorem. In fact, Birkhoff theorem assures that ϕ is a Lipschitz function and, since f and f^{-1} are smooth maps, the definition (17.7) of the homeomorphism g gives that g is bi-Lipschitz. Let $L > 0$ be the maximum between the least Lipschitz constant of g and the least Lipschitz constant of g^{-1}, i.e.

$$(17.9') \qquad L = \max\left\{\sup \frac{|g(x) - g(x')|}{|x - x'|}, \; \sup \frac{|g^{-1}(x) - g^{-1}(x')|}{|x - x'|}\right\}.$$

In view of the definition of L, we have

$$(17.9') \qquad L^{-1} \le \frac{dg(x)}{dx} \le L \text{ and } L^{-1} \le \frac{dg^{-1}(x)}{dx} \le L.$$

Let $m = \min u''$ and $M = \max u''$. They exist since u is periodic on $\mathbf{R}$. Furthermore, for the same reason, $m \le 0$. From (17.8) and (17.9) we obtain:

$$(17.10) \qquad 0 < 2L^{-1} \le 2 + m,$$

and

$$(17.10') \qquad L + L^{-1} \le 2 + M.$$

Since the function $L \to L^{-1} + L$ is monotone increasing for $L \ge 1$ and $1 \le 2/(2 + m) \le L$, as a consequence of (17.10) and $m \le 0$, the inequality (17.10') implies

$$(17.11) \qquad \frac{2 + m}{2} + \frac{2}{2 + m} \le 2 + M.$$

In the case $f = f_k$, it is not difficult to compute that $M = |k|$ and $m = -|k|$. Inequalities (17.10) and (17.11) imply

$$|k| \le 4/3.$$

This has been obtained under the assumption of the existence of a rotational invariant curve, therefore the proof is complete. $\qquad \square$

This technique was improved later by a computer aided procedure in a paper by R.S.MacKay-I.C.Percival [MK-P], who obtained that there are no rotational invariant curves in the Chirikov mapping for $|k| > 63/64$. Another result related to Theorem 17.3 has been obtained by S.Bullet [Bl]. He replaces the sine function in the definition of the Chirikov map by a piece-wise linear function and proves that for $|k| < 4/3$ rotational invariant curves may exist, for $|k| = 4/3$ *all* invariant curves appear and they eventually disappear for $|k| > 4/3$.

The variational property of rotational invariant curves.

We will prove below, following [Ma3], the basic variational property of rotational invariant curves. This property is the keystone of most of the results which will be exposed in the remaining sections, regarding in particular the destruction of invariant curves with given rotation number in §18 and the variational construction of chaotic orbits in §19.

Theorem 17.4. Let $f \in J^1$ be an exact area-preserving monotone twist C^1 diffeomorphism of the annulus and Γ a rotational invariant curve. Then every orbit of f in Γ is a minimal orbit in the sense of §14.

Proof. The rotational invariant curve Γ is the graph over $\mathbf{S}^1$ of a real valued Lipschitz function ϕ, by Birkhoff invariant curve theorem (§15). Let

$$(17.12) \qquad H(x, x') = h(x, x') - \int_x^{x'} \phi(\xi)\, d\xi \;,$$

where h is a variational principle for the lift of f to the universal cover $\mathbf{R}^2$ of the annulus. Let $g : \mathbf{S}^1 \to \mathbf{S}^1$ be the homeomorphism of the circle induced by the restriction of f on Γ, i.e.

$$(17.13) \qquad f(x, \phi(x)) = (g(x), \phi(g(x))) \;.$$

As previously remarked, g is a bi-Lipschitz homeomorphism, since ϕ is a Lipschitz function and f, f^{-1} are C^1 maps. Furthermore g is orientation preserving, since f is orientation preserving and also preserves both ends of the annulus (infinite cylinder). At a point (x, x'), where $x' = g(x)$, the following holds:

$$(17.14) \quad H_1(x, x') = h_1(x, x') + \phi(x) = 0 \quad \text{and} \quad H_2(x, x') = h_2(x, x') - \phi(x') = 0 \;.$$

This follows from (17.13) and from the generating equations (5.2) for f, i.e. $f(x, y) = (x', y')$ if and only if $y = -h_1(x, x')$ and $y' = h_2(x, x')$. Consequently,

$$(17.15) \qquad \frac{d}{dx} H(x, g(x)) = H_1(x, g(x)) + H_2(x, g(x))\frac{dg}{dx} = 0 \;,$$

and this equation holds almost everywhere, since the derivative dg/dx is defined almost everywhere (and positive), being g a Lipschitz increasing function. Therefore, $H(x, g(x))$, being a Lipschitz function whose derivative vanishes almost everywhere, is constant equal to $C \in \mathbf{R}$.

Since $H_{12}(x, x') = h_{12}(x, x') < 0$, and $H_1(x, g(x)) = H_2(x, g(x)) = 0$, it follows that
$$(17.16)$$
$$H_1(x, x') < 0 \;, \quad \text{when} \quad x' > g(x) \quad \text{and} \quad H_1(x, x') > 0 \quad \text{when} \quad x' < g(x) \;,$$
$$H_2(x, x') < 0 \;, \quad \text{when} \quad x' < g(x) \quad \text{and} \quad H_2(x, x') > 0 \quad \text{when} \quad x' > g(x) \;.$$

Consequently, $H(x, x') > C$, when $x' \neq g(x)$ (and $H(x, x') = C$, when $x' = g(x)$).

Suppose $\mathcal{O} = \{(x_i, y_i)\}_{i \in \mathbf{Z}}$ represents an orbit of the diffeomorphism f on the rotational invariant curve Γ. By (17.13), the stationary configuration $x = (x_i)_{i \in \mathbf{Z}}$ associated to the orbit $\mathcal{O}$ (see §3 and §14) satisfies $g(x_i) = x_{i+1}$. Let $m < n$ be integers and let $y = (y_m, ..., y_n)$ be a segment of a configuration subject to the constraint $y_m = x_m$ and $y_n = x_n$. Suppose that $y_i \neq x_i$ for some i, $m < i < n$.

Then

(17.17)

$$h_{mn}(x) = \sum_{i=m}^{n-1} H(x_i, x_{i+1}) + \int_{x_m}^{x_n} \phi(\xi)\,d\xi = (n-m)C +$$

$$+ \int_{x_m}^{x_n} \phi(\xi)\,d\xi < \sum_{i=m}^{n-1} H(y_i, y_{i+1}) + \int_{x_m}^{x_n} \phi(\xi)\,d\xi = h_{mn}(y)\,,$$

where $h_{mn} : \mathbf{R}^{n-m} \to \mathbf{R}$ is the function

(17.18)
$$h_{mn}(y) = \sum_{i=m}^{n-1} h(y_i, y_{i+1})\,.$$

Thus, we have shown that x is a minimal configuration in the sense of §9 and, consequently, $\mathcal{O}$ is a minimal orbit, according to the definition given in §14. □

§18. Destruction of Invariant Curves.

The variational property of rotational invariant curves, described at the end of §17, explains the relevance in the context of area-preserving monotone twist mappings of a notion of a barrier function, originally introduced in solid state physics in connection with the Frenkel-Kontorova model mentioned in §3 [Au-LeD]. This barrier function provides a tool which is sensitive to the existence of a rotational invariant curve of given rotation number.

The Peierls's barrier.

The Peierl's barrier is a real valued function depending on a rotation symbol ω (defined in §13) and on $\xi \in \mathbf{R}$, where $\mathbf{R}$ is seen as the universal cover of the circle $\mathbf{S}^1$. It measures to which extent the stationary configuration $(y_i)_{i \in \mathbf{Z}}$, subject to the condition $y_0 = \xi$, is not minimal. In §13, we introduced the quantity (13.3)

$$\Delta h(y, x) = \sum_{-\infty}^{+\infty} h(y_i, y_{i+1}) - h(x_i, x_{i+1}) ,$$

for the study of minimal configurations of rotation symbol $p/q^\pm$. To include the case of rational numbers as rotation symbols, we introduce the quantity $\Delta_\omega h(y, x)$, as follows:

$$(18.1) \qquad \Delta_\omega h(y, x) = \sum_I h(y_i, y_{i+1}) - h(x_i, x_{i+1}) ,$$

where I is equal to $\mathbf{Z}$ when ω is an irrational rotation symbol or is equal to $p/q^\pm$ and it is equal to $\{0, ..., q-1\}$, when $\omega = p/q$. The above quantity can be shown to be convergent (possibly to $+\infty$) whenever x is a minimal configuration of rotation symbol ω and the configuration y is asymptotic to x (i.e. $|y_i - x_i| \to 0$ as $i \to \pm\infty$). The proof of this fact is contained in Theorem 13.1, in the case $\omega = p/q^\pm$, but it works as well in the general case. We recall that, for any rotation symbol ω, we introduced in §12-13 A_ω as the subset of the real line which is the union of all minimal configurations of rotation symbol ω. In Theorem 12.2 and 13.4 we proved that A_ω is a closed subset of $\mathbf{R}$, for any rotation symbol ω. The *Peierls's barrier* is defined as follows. Let ω be a rotation symbol and $\xi \in \mathbf{R}$, then $P_\omega(\xi) = 0$, if $\xi \in A_\omega$. In the case $\xi \notin A_\omega$, hence it belongs to a complementary interval $J = (J^-, J^+)$ to A_ω (i.e. a connected component of $\mathbf{R} \setminus A_\omega$), $P_\omega(\xi)$ is defined as

$$(18.2) \qquad P_\omega(\xi) = \min \{\Delta_\omega h(y, x^-) \,|\, y_0 = \xi\} ,$$

where the minimum is taken over the set of all configurations satisfying $x_i^- \leq y_i \leq x_i^+$, for all $i \in \mathbf{Z}$, and $x^\pm$ are the minimal configurations of rotation symbol ω satisfying $x_0^\pm = J^\pm$ (which certainly exist being A_ω closed). In the case $\omega = p/q$, the minimum is taken under the additional periodicity constraint $y_{i+q} = y_i + p$. We recall that, under the above constraints, the quantity $\Delta_\omega h(y, x^-)$ is finite, as a consequence of (12.2) (which holds true if $\omega = p/q^\pm$), in the case $\omega \neq p/q$. Furthermore, the minimum in (18.2) exists since the topological space given by the infinite product of the intervals $[x_i^-, x_i^+]$, $i \in \mathbf{Z}$, is compact, with respect to the product topology, and $\Delta_\omega h(y, x^-)$ is a continuous function of y. In the case $\omega = p/q$

there is no difficulty related to the finiteness of $\Delta_\omega h(y, x^-)$ or to the existence of the minimum in (18.2). In the definition of the Peierls's barrier, we may take the minimum over the larger set of configurations y of rotation symbol ω, subject to the constraint $y_0 = \xi$, for which the quantity $\Delta_\omega h(y, x^-)$ is defined. In fact, any such configuration can be replaced by one satisfying in addition the condition $x_i^- \le y_i \le x_i^+$ decreasing $\Delta_\omega h(y, x^-)$. This is a consequence of the minimality of $x^\pm$ and of the inequality (9.1) (Aubry crossing lemma). It is sufficient to consider instead of y the configuration y' defined as $y' = x_- \vee y \wedge x^+$.

Lemma 18.1. $P_\omega(\xi) \ge 0$, for any $\xi \in \mathbf{R}$, and $P_\omega(\xi) = 0$ if and only if $\xi \in A_\omega$. In particular, if $\omega \in \mathbf{R} \setminus \mathbf{Q}$, $P_\omega(\xi) \equiv 0$ if and only if there exists a rotational invariant curve of rotation number ω.

Proof. It is part of the definition that $P_\omega(\xi) = 0$ if $\xi \in A_\omega$. If $\xi \notin A_\omega$, suppose $\xi \in J = (J^-, J^+)$, where J is a complementary interval to A_ω. Let $x^\pm$ be minimal configurations of rotation symbol ω such that $x_0^\pm = J^\pm$. Suppose that y is a configuration satisfying $y_0 = \xi$, $x_i^- \le y_i \le x_i^+$, for all $i \in \mathbf{Z}$, and $y_{i+q} = y_i + p$, in case $\omega = p/q$. Then, $\Delta_\omega(y, x^-) > 0$. Furthermore, $P_\omega(\xi)$ is given by (18.2) for all $\xi \in [J^-, J^+]$. In the case that $\omega = p/q$, we already proved in Theorem 10.3 that if y is a periodic configuration of type (p, q) and minimizes

$$h_q^{per}(x) = \sum_{i=a}^{a+q-1} h(x_i, x_{i+1})$$

over all configurations of type (p, q), then it is minimal. This clearly implies the above assertions in case $\omega = p/q$, since the configuration y is not minimal and the configurations $x^\pm$ are minimal. When ω is not a rational number, x^- and x^+ (and therefore y) are asymptotic, by (12.2), and $\omega - 2 \le |x_{i+1}^\pm - x_i^\pm| \le \omega + 2$, by (12.4). Therefore $\Delta_\omega(y, x^-)$ is convergent and finite. Since y is not minimal, there exists a configuration v, asymptotic to $x^\pm$, such that

$$\Delta_\omega(y, x^-) > \Delta_\omega(v, x^-) .$$

Let $m < n$ be integers and consider the segment of a configuration $v^{m,n}$ defined as

$$v^{m,n} = \begin{cases} v_i & \text{if } i \ne m, n , \\ x_i^- & \text{if } i = m, n . \end{cases}$$

Since h satisfies property (H_1) and it is continuous, it is uniformly continuous in the region defined by $\omega - 2 \le |x_{i+1}^\pm - x_i^\pm| \le \omega + 2$. Furthermore the series defining $\Delta_\omega(v, x^-)$ is convergent. Therefore, for any $\epsilon > 0$ it is possible to choose $N \in \mathbf{N}$ so that, for $|n|, |m| \ge N$,

$$(18.3) \qquad |\Delta_\omega(v, x^-) - \sum_{i=m}^{n-1} h(v_i^{m,n}, v_{i+1}^{m,n}) - h(x_i^-, x_{i+1}^-)| < \epsilon .$$

Since x^- is a minimal configuration, (18.3) clearly implies $\Delta_\omega(v, x^-) \ge 0$. Thus, $\Delta_\omega(y, x^-) > 0$. Furthermore, since $x^\pm$ are minimal and asymptotic, it follows easily that $\Delta_\omega(x^+, x^-) = \Delta(x^-, x^-) = 0$, thereby proving the above remark that $P_\omega(\xi)$ is given by (18.2) for all $\xi \in [J^-, J^+]$. The assertions we just proved immediately imply the first part of the statement of Lemma 18.1. On the other hand, there exists a rotational invariant curve of rotation number $\omega \in \mathbf{R} \setminus \mathbf{Q}$ if and only if $A_\omega = \mathbf{R}$. This follows from Theorem 14.1 and Theorem 17.4. Therefore the argument is completed. $\qquad\qquad\qquad\qquad\qquad\qquad\qquad\qquad\qquad\qquad\qquad\qquad\qquad$ □

The dependence of the Peierls's barrier with respect to $\xi \in \mathbf{R}$ is described in [Ma8, Lemma 6.3]: if h is a variational principle wich satisfies the conditions (H_θ), introduced in §7, then $P_\omega(\xi)$ is a Lipschitz function of $\xi \in \mathbf{R}$, whose Lipschitz constant can be taken to be 2θ. The argument applies estimates on the derivatives of the variational principle h, which are a consequence of (H_θ), and the Aubry Crossing Lemma.

The fundamental fact about the Peierls's barrier is that it satisfies a modulus of continuity with respect to the rotation symbol. The proof of this property, contained in [Ma8]-[Ma9], is rather technical and consists in a delicate comparison, heavily relying on Aubry theory of minimal configurations exposed in §§9-13, between the "energy" of minimal configurations of rational and irrational rotation symbols. The result is the following:

Theorem 18.2. *There exists a positive real number C such that the following holds. Suppose h satisfies the conditions (H_θ). If p/q is a rational number (in lowest terms) and ω is a rotation symbol, then*

a) $|P_\omega(\xi) - P_{p/q}(\xi)| \leq C\theta(q^{-1} + |q\pi(\omega) - p|)$

b) $|P_\omega(\xi) - P_{p/q^+}(\xi)| \leq C\theta|q\pi(\omega) - p|$ *in the case* $\omega \geq p/q^+$ *, and*

$|P_\omega(\xi) - P_{p/q^-}(\xi)| \leq C\theta|q\pi(\omega) - p|$ *in the case* $\omega \leq p/q^-$ *,*

where $\pi(\omega)$ denotes the underlying number associated to the rotation symbol ω (see §13).

Furthermore, in [F], it has been shown that, under certain conditions on the Peierls's barrier $P_{p/q}(\xi)$, where p/q is a rational number in lowest terms, related to the hyperbolicity of the minimal periodic orbits of type (p, q), the modulus of continuity in part b) of Theorem 18.2 can be replaced by an exponential modulus of continuity. The proof combines the methods of [Ma8-9] with Herman's approach to the destruction of invariant circles in the completely integrable case [He]. The result is essentially the following, although some further technical difficulties are not discussed.

Theorem 18.3. *There exist constants C and $C(\theta)$ (depending only on $\theta > 0$) such that the following holds. Given a variational principle h, satisfying the conditions (H_θ), and a rational number p/q (in lowest terms), assume that:*

a) *there is a single minimal periodic orbit of type (p, q);*

b) *there exists $\lambda_{(p,q)} > 0$ such that $P_{p/q}(\xi) \geq \lambda_{(p,q)} \operatorname{dist}(\xi, A_{p/q})^2$, for any $\xi \in \mathbf{R}$, where $\operatorname{dist}(\cdot, A_{p/q})$ denotes the euclidean distance function from the closed set $A_{p/q}$ on the real line.*

Then, for any rotation symbol ω whose underlying number $\pi(\omega)$ is irrational and satisfies $|q\pi(\omega) - p| \leq \Omega_{(p,q)}$,

$$|P_\omega(\xi) - P_{p/q^\pm}(\xi)| \leq C(\theta) \exp\left(-\Omega_{(p,q)}/|q\pi(\omega) - p|\right),$$

$+$ *or* $-$ *sign according to whether $\omega > p/q$ or $\omega < p/q$, where*

$$\Omega_{(p,q)} = C/q(1 + \lambda_{(p,q)}^{-1/2}).$$

The conditions a) and b) in Theorem 18.3 are generic in any smooth topology and in the analytic topology. However this result can be applied to the destruction of invariant curves only when it is possible to achieve a good lower bound for the constant λ, which is related to the *hyperbolicity* of the (unique) minimal periodic orbit of type (p, q). This has been done [F] in the case of the completely integrable map (17.6) and it is conjectured to be possible also in the case of a rotational invariant curve γ such that the restriction $f|\gamma$ of the twist diffeomorphism f to γ is sufficiently smoothly conjugate to a rigid rotation. In the general case the problem of constructing a perturbation which yields the desired lower bounds for λ in Theorem 18.3 is open.

In the following we will apply the reduction of a periodic (minimal) orbit to a fixed (minimal) point. This is usually done in the theory of dynamical systems by considering compositions of the map with itself. In the case of exact area-preserving monotone twist mappings, generated by a variational principle, composition corresponds at the level of variational principles to the operation of *conjunction*, introduced in [Ma8]. Given two variational principles h_1 and h_2, satisfying the condition (H_2), their conjunction is defined in the following way:

$$(18.4) \qquad (h_1 \star h_2)(x, x') = \min_{\xi \in \mathbf{R}}\left(h_1(x, \xi) + h_2(\xi, x')\right)$$

(the minimum exists by condition (H_2)).

In [Ma8] it is proved that, if both h_1 and h_2 satisfy the conditions (H_θ), then so does the conjunction $h_1 \star h_2$ with the same θ. Notice that, even when both h_1 and h_2 are smooth, the conjunction $h_1 \star h_2$ needs not to be smooth.

Given a variational principle h and a rational number p/q (in lowest terms), we will often consider in the following the variational principle H (depending on p/q) defined by

$$(18.5) \qquad H(x, x') = (h \star \ldots \star h)(x, x' + p)(q - \text{times}) + \text{constant},$$

where the additive constant will be chosen so that $\min H(x, x) = 0$. One has

$$(18.5') \qquad A_\omega^h = A_{q\omega - p}^H \quad \text{and} \quad P_\omega^h(\xi) = P_{q\omega - p}^H(\xi),$$

as long as ω is a rotation symbol which is not a rational number or is a rational number whose denominator is divisible by q.

The destruction of invariant curves.

The qualitative principle underlying the destruction results for invariant curves can be described as follows. It is known since the work of Poincaré that rotational invariant curves of rational rotation number can be destroyed by a perturbation as small as we wish in any smooth topology or in the analytic topology. On the other hand, it follows from Birkhoff invariant curve theorem, namely from the fact that rotational invariant curves are Lipschitz graphs whose Lipschitz constant is determined a priori from the map, that the set of rotational invariant curves is closed. Therefore, the destruction of an invariant curve with rational rotation number p/q is accompanied by the destruction of an open set of nearby curves, including those with irrational rotation numbers contained in a certain open interval $I(p, q)$ around p/q . To obtain a destruction theorem for invariant curves of a *fixed* irrational rotation number, one has to provide lower *estimates* for the size of the interval $I(p,q)$, in terms of (p, q) (in fact in terms only of the denominator $q > 0$). In view of Lemma 18.1, this can be done using the modulus of continuity for the Peierls's barrier.

The destruction of invariant curves of rational rotation number p/q is described, in terms of the Peierls's barriers $P_{p/q}(\xi)$ and $P_{p/q\pm}(\xi)$, in the following. Details can be found, in the smooth case, in [Ma9], and in the analytic case, in [F].

Lemma 18.4. *There exists positive constants $C_r(\theta)$ such that the following holds. Given a smooth variational principle h_f, satisfying the conditions (H_θ), associated to an exact area-preserving monotone twist mapping f, for any $r \in N$ and $\epsilon > 0$, there exists a periodic smooth function w on $\mathbf{R}$ such that:*

$$\|w\|_{r+1} \leq \epsilon \quad \text{and} \quad \max_{\xi \in \mathbf{R}} P^{h_g}_{p/q\pm}(\xi) \geq C_r(\theta)\, \epsilon^{r+2}/q^{(r+1)^2} ,$$

where $h_g(x, x') = h_f(x, x') + w(x)$ and $\|\cdot\|_r$ denotes the C^r norm on the space of real-valued functions on S^1.

Proof. Choose a minimal configuration $x = (x_i)_{i \in \mathbf{Z}}$ of type (p, q). Then the set $\{x_i + j \mid i, j \in \mathbf{Z}\}$ intersects each interval $[a, a+1)$, $a \in \mathbf{R}$, in exactly q points. By the pidgeon hole principle, there exists a complementary interval J of length $\geq 1/q$ to this set. We choose a smooth non-negative function u, satisfying $u(x+1) = u(x)$, supported in $J + i$, $i \in \mathbf{Z}$, whose C^{r+1} norm is small. If $\|u\|_{r+1} \leq \epsilon/2$, then, since the length of J is $\geq 1/q$, u can be chosen such that

$$(18.6) \qquad\qquad \max u \geq C_r\, \epsilon/q^{r+1} .$$

On the other hand, since the support of u does not contain any point of the minimal configuration $x = (x_i)_{i \in \mathbf{Z}}$ and u is non-negative,

$$(18.7) \qquad\qquad P^{h_u}_{p/q}(\xi) = P^{h_f}_{p/q}(\xi) + u(\xi) ,$$

where h_u is the variational principle $h_u(x, x') = h_f(x, x') + u(x)$. This follows immediately from the definition of the Peierls's barrier $P_{p/q}$. We will now use the reduction to the case of a (minimal) fixed point. Let H_u be the variational principle associated to h_u as in (18.5). It follows from (18.5) and (18.5') that

$$(18.8) \qquad\qquad H_u(\xi, \xi) = P^{H_u}_0(\xi) = P^{h_u}_{p/q}(\xi) ,$$

which, in view of (18.7) implies

$$(18.8') \qquad\qquad H_u(\xi, \xi) \geq u(\xi) .$$

Let $J = (J^-, J^+)$ be the complementary interval to the set $\{x_i + j \mid i, j \in \mathbf{Z}\}$ containing the support of u. By a result in the Aubry theory of minimal configurations (Corollary 13.6), since, by (18.5'), $A_0^{H_u} = A_{p/q}^{h_u}$, there exist minimal configurations $x^\pm$ of rotation symbol $0^\pm$ such that $x_i^- \to J^\pm$ as $i \to \mp\infty$ and $x_i^+ \to J^\pm$ as $i \to \pm\infty$. We are interested in a lower bound for the maximum, over $i \in \mathbf{Z}$, of $|x_{i+1}^\pm - x_i^\pm|$. Let J' be the middle third of the interval J. If no $x_i^\pm \in J'$, then we may take the length of J' as a lower bound. Suppose $x_i^\pm \in J'$. Removing $x_i^\pm$ from the configuration $x^\pm$, we obtain a new configuration $y^\pm$ of rotation symbol $0^\pm$, namely $y_j^\pm = x_j^\pm$, for $j < i$, and $y_j^\pm = x_{j+1}^\pm$, for $j \geq i$. By Theorem 13.1, the quantity (13.3)

$$\Delta(y^\pm, x^\pm) = \sum_{i=-\infty}^{+\infty} \left(H_u(y_i^\pm, y_{i+1}^\pm) - H_u(x_i^\pm, x_{i+1}^\pm) \right)$$

exists and it is positive. On the other hand, clearly

$$(18.9) \qquad \Delta(y^\pm, x^\pm) = H_u(x_{i-1}^\pm, x_{i+i}^\pm) - H_u(x_{i-1}^\pm, x_i^\pm) - H_u(x_i^\pm, x_{i+i}^\pm) .$$

Since H_u satisfies the conditions $(H_{\theta'})$, for $\theta' = \theta + 1$, provided that ϵ is sufficiently small, the following estimate holds [Ma9, §3]:

$$(18.9')$$
$$H_u(x_{i-1}^\pm, x_{i+i}^\pm) - H_u(x_{i-1}^\pm, x_i^\pm) - H_u(x_i^\pm, x_{i+i}^\pm) \leq \theta' \, (x_{i+1}^\pm - x_{i-1}^\pm) - H_u(x_i, x_i) .$$

Therefore, by the positivity of $\Delta(y^\pm, x^\pm)$ and, since $H_u(\xi, \xi) \geq u(\xi) \geq C_r' \epsilon/q^{r+1}$ on J', as a consequence of (18.6) and (18.8'), and $x_i^\pm \in J'$, there exists $i \in \mathbf{Z}$ such that

$$(18.10) \qquad\qquad |x_{i+1}^\pm - x_i^\pm| \geq C_r(\theta) \, \epsilon/2q^{r+1} .$$

Let v be a non-negative smooth function on $\mathbf{R}$, satisfying $v(x + 1) = v(x)$, whose support is contained in $[x_i^\pm, x_{i+1}^\pm] + j$, $j \in \mathbf{Z}$. Since the estimate (18.10) on the length of $[x_i^\pm, x_{i+1}^\pm]$ holds, if $\|v\|_{r+1} \leq \epsilon/2$, it is possible to choose v satisfying

$$(18.11) \qquad \max v \geq C_r \epsilon (C_r(\theta) \, \epsilon/q^{r+1})^{r+1} = C_r(\theta) \, \epsilon^{r+2}/q^{(r+1)^2} .$$

Since v vanishes outside the union of the intervals $[x_i^\pm, x_{i+1}^\pm] + j$, $j \in \mathbf{Z}$, and it is non-negative,

$$(18.12) \qquad\qquad P_{p/q^\pm}^{h_g}(\xi) \geq P_{p/q^\pm}^{h_u}(\xi) + v(\xi) .$$

where $w = u + v$ and $h_g(x, x') = h_f(x, x') + w(x) = h_u(x, x') + v(x)$. The inequalities (18.11) and (18.12) immediately give the desired estimate. $\qquad\square$

In the analytic case, the argument is slightly complicated because of the absence of compactly supported analytic functions. The appropriate substitutive tools are a version of the maximum principle for holomorphic functions (Hadamard 3-circle theorem) and the approximation theorem for real analytic functions by trigonometric polynomials (Jackson approximation theorem). The outline of the argument is, in any other respect, similar to the smooth case. The result is the following [F]:

Lemma 18.5. *There exist positive constants C, $C(\theta)$ such that the following holds. Given a real analytic variational principle h_f, satisfying the conditions (H_θ), for any $r > 0$ and $\epsilon > 0$ there exists a trigonometric polynomial w such that:*

$$|w|_r \leq \epsilon \quad \text{and} \quad \max_{\xi \in \mathbf{R}} P^{h_g}_{p/q\pm}(\xi) \geq (\epsilon/4) \exp\left(-C(\theta)\,\epsilon_1^{-3/2}\,\exp(Crq)\right) ,$$

where $h_g(x, x') = h_f(x, x') + w(x)$, $|w|_r$ denotes the maximum modulus on the infinite strip $S_r = \{z \in \mathbf{C} \mid |\operatorname{Im} z| \leq r\}$ of the holomorphic extension of the trigonometric polynomial w to the complex plane $\mathbf{C}$, and $\epsilon_1 = \min(\epsilon, 1)$.

Combining Lemma 18.4 and 18.5 with the modulus of continuity for the Peierls's barrier, we obtain destruction results for invariant curves.

Theorem 18.6. *[Ma9] Let γ be a rotational invariant curve of rotation number $\rho(\gamma) = \omega \in \mathbf{R} \setminus \mathbf{Q}$, for an exact area-preserving monotone twist C^∞ diffeomorphism $f \in J^\infty_*$ of the annulus. Assume ω is a Liouville number, i.e. the Diophantine condition (16.1) does not hold. Then in any neighborhood $\mathcal{U}_f$ of f in the C^∞ topology there exists an exact area-preserving monotone twist diffeomorphism $g \in J^\infty_*$ which does not admit any invariant circle of rotation number ω.*

Proof. We argue by contradiction. Assume that there exists a neighbourhood $\mathcal{U}_f$ of f in the C^r topology such that any $g \in \mathcal{U}_f$ has a rotational invariant curve of rotation number ω. In particular, if $\epsilon > 0$ is sufficiently small, the diffeomorphism g associated to the variational principle h_g, constructed in Lemma 18.4, belongs to $\mathcal{U}_f$. Therefore, g has a rotational invariant curve of rotation number ω and, by Lemma 18.1,

$$P^{h_g}_\omega(\xi) \equiv 0 ,$$

which implies, by Theorem 18.2 and Lemma 18.4,

(18.13)
$$C_r(\theta)\,\epsilon^{r+2}/q^{(r+1)^2} \leq C\theta|q\omega - p| ,$$

and this holds for any $p, q \in \mathbf{Z}$, $q > 0$. Therefore ω satisfies a Diophantine condition (16.1) with exponent $(r + 1)^2$ and we reached the desired contradiction. $\qquad\square$

In fact the argument given in the proof of Theorem 18.6 proves that a necessary condition for the stability of a rotational invariant curve of rotation number ω under perturbations sufficiently small in the C^r topology is that ω is Diophantine with exponent $\geq (r+1)^2$. On the other hand, the K.A.M theorem (Theorem 16.1) asserts that the curve is in fact stable under C^r small perturbations, if the Diophantine exponent of its rotation number is $\leq (r-3)/2$. Therefore, the stability and destruction results for invariant curves in the C^r's topologies exhibit a gap which does not appear in the C^∞ case. Namely, the dependence of the Diophantine exponent of the rotation number of the invariant curve is $O(r)$ in the K.A.M theorem and $O(r^2)$ in Theorem 18.6, as $r \to \infty$. This gap leaves open the possibility of non-KAM stability phenomena. We will see that the gap is even more significant in the analytic topology.

In the particular case of the *completely integrable map* (17.6), the modulus of continuity given by Theorem 18.3 can be applied, thereby obtaining a destruction result, due to Herman [He], where the gap disappears and the dependence is $O(r)$ as in the K.A.M. theorem. In fact, any stationary configuration $(x_i)_{\mathbf{Z}}$ of type (p, q) in the completely intergrable map is *equispaced* mod. $\mathbf{Z}$, i.e. the set $\{x_i + j \,|\, i, j \in \mathbf{Z}\}$ intersects any interval $[a, a+1)$, $a \in \mathbf{R}$, in q points exactly equispaced. Therefore, it is possible to choose the function u which gives the first step in the construction of the perturbation in Lemma 1.4 such that $\|u\|_{r+1} \leq \epsilon/2$ and

$$(18.14) \qquad u''(x_i) \geq C_r \epsilon / q^{r-1} \,,$$

besides satisfying (18.6). To this purpose, it would be sufficient that the complementary set to $\{x_i + j \,|\, i, j \in \mathbf{Z}\}$ contains two *adjacent* intervals of length $\geq C/q$, where $C > 0$ is a universal constant independent of $q > 0$. Since $w''(x_i) = u''(x_i)$, $i \in \mathbf{Z}$, and

$$P^{h_g}_{p/q}(\xi) = P^{h_f}_{p/q}(\xi) + w(\xi) \,,$$

the constant $\lambda_{(p,q)} > 0$ contained in Theorem 18.3 can be estimated as follows:

$$(18.15) \qquad \lambda_{(p,q)} \geq C_r \epsilon / q^{r-1} \,.$$

This holds in view of the fact that $P^{h_f}_{p/q}(\xi) \equiv 0$ in the case f is the completely integrable map. Then Theorem 18.3, together with Lemma 1.4, allow to replace the estimate (18.13) in the proof of Theorem 18.6 by the following:

$$(18.16) \qquad C_r(\theta)\, \epsilon^{r+2}/q^{(r+1)^2} \leq C(\theta) \exp\left(-C_r \epsilon^{1/2} q^{-(r+1)/2} |q\omega - p|^{-1}\right) \,,$$

which leads to

$$(18.16') \qquad |q\omega - p| \geq C_r(\theta, \epsilon)/q^{(r+1)/2} \operatorname{Log} q \,,$$

for any $p, q \in \mathbf{Z}$, $q > 0$. Therefore, in this case, we obtain a destruction result in the C^r topology under the hypothesis that the Diophantine exponent of the rotation number ω is $> (r+1)/2$. This has to be compared with the condition given in the K.A.M. theorem, according to which we have stability in the C^r topology provided that the Diophantine exponent is $\leq (r-3)/2$. We see that there is no gap in the behaviour as $r \to +\infty$. In the general case the estimate (18.15) is not available, as a consequence of the fact that, if a (minimal) periodic orbit is not equispaced mod. **Z**, it is unclear whether it is possible to produce a sufficiently hyperbolic orbit (with estimates) by a perturbative construction.

In the analytic case, similar arguments, relying on Lemma 18.5 instead of Lemma 18.4, lead to the following results contained in [F]:

Theorem 18.7. *Let γ be a rotational invariant curve of rotation number $\rho(\gamma) = \omega \in \mathbf{R} \setminus \mathbf{Q}$, for an exact area-preserving monotone twist analytic diffeomorphism f of the annulus. Assume ω satisfies the condition*

$$(I) \qquad \limsup_{n \to +\infty} \frac{\operatorname{Log} \operatorname{Log} q_{n+1}}{q_n} > 0 \, ,$$

where $(p_n/q_n)_{n \in \mathbf{N}}$ is the sequence of convergents of the continued fraction expansion of ω. Then in any neighborhood $\mathcal{U}_f$ of f in the analytic topology there exists an exact area-preserving monotone twist analytic diffeomorphism g which does not admit any rotational invariant curve of rotation number ω.

Proof. Let $U_{r,\epsilon}$ be the set of all real analytic periodic functions w on $\mathbf{R}$ which extends to a holomorphic function W on the strip $S_r = \{z \in \mathbf{C} \,|\, |\operatorname{Im} z| \leq r\}$, in the complex plane $\mathbf{C}$, and $|W|_r < \epsilon$, where, as before, $|W|_r$ denotes the maximum of $|W|$ on S_r. For any function $\epsilon : \mathbf{R}^+ \to \mathbf{R}^+$, let

$$(18.17) \qquad \mathcal{U}_\epsilon =: \bigcup_{r > 0} U_{r, \epsilon(r)} \, .$$

The family of sets described in (18.17), as the function ϵ varies, is a basis of open sets for the analytic topology. Assume by contradiction that there exists a strictly positive function $\epsilon : \mathbf{R}^+ \to \mathbf{R}^+$ such that, whenever $h_f - h_g \in \mathcal{U}_\epsilon$, then g admits a rotational invariant curve of rotation number ω and therefore

$$(18.18) \qquad P_\omega^{h_g}(\xi) \equiv 0 \, .$$

Let $(p_n/q_n)_{n \in \mathbf{N}}$ be the sequence of convergents of the continued fraction expansion of ω. Then, given any $r > 0$ and any $\epsilon < \epsilon(r)$, by Lemma 18.5 one can construct, for each $n \in \mathbf{N}$, an exact area-preserving monotone twist analytic diffeomorphism g_n such that $h_f - h_n \in U_{r,\epsilon} \subset \mathcal{U}_\epsilon$, where h_n is a variational principle for g_n and

$$(18.19) \qquad \max_{\xi \in \mathbf{R}} P_{p_n/q_n}^{h_n}{}_{,\pm}(\xi) \geq (\epsilon/4) \exp\left(-C(\theta)\epsilon_1^{-3/2} \exp(Crq_n)\right) \, .$$

By the modulus of continuity given in Theorem 18.2:

$$(18.20) \qquad (\epsilon/4) \exp\left(-C'(\theta)\epsilon_1^{-3/2} \exp(Crq_n)\right) \leq C\theta|q_n\omega - p_n| \leq C\theta/q_{n+1} \, ,$$

where the last inequality holds by the known properties of the continued fraction expansion.

By (18.20), taking the logarithm, one gets

(18.21)
$$\text{Log } q_{n+1} \le C_0(\theta) + \text{Log}(\epsilon_1^{-1}) + C(\theta)\epsilon_1^{-3/2} \exp(Crq_n) \le$$
$$\le C_1(\theta)\epsilon_1^{-3/2} \exp(Crq_n) \ ,$$

and, by taking the logarithm again in (18.21),

(18.21')
$$\text{Log Log } q_{n+1} \le \text{Log } C_1(\theta) - 3/2 \text{ Log } \epsilon_1 + Crq_n \ ,$$

which, since $q_n \ge 2^n$, leads to

(18.22)
$$\limsup_{n \to +\infty} \frac{\text{Log Log } q_{n+1}}{q_n} \le Cr \ .$$

Since $r > 0$ was arbitrarily chosen, (18.22) implies that condition (I) is not satisfied, which contradicts the initial assumption and concludes the proof of the theorem. $\square$

In the completely integrable case, the double logarithm in condition (I) is replaced by the single logarithm of condition (J) since, in applying the exponential modulus of continuity for the Peierls's barrier (Theorem 18.3), one of the two exponentials appearing in the estimate provided by Lemma 18.5 cancels out. In fact, it is possible to construct a small analytic perturbation, given by a trigonometric polynomial w satisfying $|w|_r \le \epsilon$, thereby achieving:

(18.23)
$$\lambda_{(p,q)} \ge C\epsilon_1 \, q^2 \exp(-r \, q) \ .$$

The estimate (18.23) should be compared with (18.15) which holds in the smooth case. The estimate (18.20) is then replaced by the following:

(18.24)
$$(\epsilon/4)\exp\left(-C(\theta)\epsilon_1^{-3/2} \exp(Cr \, q_n)\right) \le$$
$$\le C(\theta)\exp\left(C'\epsilon_1^{1/2} \exp(-r \, q_n/2)|q_n\omega - p_n|^{-1}\right) \ ,$$

by applying the exponential modulus of continuity given in Theorem 18.2 and Lemma 18.5. Taking twice the logarithm in (18.24), it is not difficult to obtain:

(18.25)
$$\limsup_{n \to +\infty} \frac{\text{Log } q_{n+1}}{q_n} \le C''r \ .$$

Therefore, the following holds:

Theorem 18.8. *Under the same hypotheses of Theorem 18.7, assume furthermore that γ is a rotational invariant curve of the completely integrable map* (17.6). *Then, if its rotation number ω satisfies the condition*

(J)
$$\limsup_{n \to +\infty} \frac{\text{Log } q_{n+1}}{q_n} > 0 \ ,$$

the same conclusion as in Theorem 18.7 holds.

Theorem 18.8 provides a partial converse to the analytic K.A.M. theorem due to Rüssmann, mentioned in §16. The gap between the Brjuno condition (16.2) of

Rüssmann's theorem and the above condition (J) is probably due to limits of the destruction method employed. More significant appears the gap between the conditions (I) and (J). These conditions are related to analogous conditions obtained for the problem of linearizing a holomorphic germ, having an elliptic fixed point with multiplier $\exp(2\pi i\omega)$ (Siegel center problem). This problem has been studied for many years as a simplified model for the "small divisors" problems which arise in the stability problem for invariant tori of real analytic hamiltonian systems. Thanks to the work of Yoccoz and Perez-Marco the picture in the case of the Siegel problem is fairly complete. We refer to the survey article [P-M]. The Rüssmann's K.A.M. theorem together with the above Theorems 18.7 and 18.8 suggest a similar picture for the case of rotational invariant curves of exact area-preserving monotone twist mappings, but here the description is incomplete and difficult problems concerning especially the behaviour of non-smooth curves are open.

§19. Dynamics in the Sthocastic Regions.

As mentioned above, Birkhoff invariant curve theorem implies that the union of the set of all rotational invariant curves is a closed subset of the annulus. A connected component of its complementary set, which is open and homemorphic to an annulus, is called a *Birkhoff region of instability*. This is characterized by the fact that it contains no rotational invariant curves. Nevertheless, for any rotation number $\omega \in \mathbf{R} \setminus \mathbf{Q}$, there exists an action minimizing invariant set Σ_ω, described in §14, which is given by the union of all minimal orbits of rotation symbol ω, replacing the invariant curve of rotation number ω. The Aubry-Mather set $\Sigma_\omega^\star$ is the unique minimal set (in the sense of topological dynamics) contained in the set Σ_ω. Furthermore, $\Sigma_\omega^\star$ is a *Denjoy* invariant set, i.e. from the point of view of ergodic theory, the dynamics of the map on $\Sigma_\omega^\star$ is isomorphic to the dynamics of the rigid rotation R_ω of the circle. This also implies that $\Sigma_\omega^\star$ is uniquely ergodic, i.e. it supports a unique Borel probability invariant measure σ_ω. In [Ma6], the first author has shown that, when the minimizing set Σ_ω is not an invariant curve, as it is the case in a Birkhoff region of instability, then an abundance of Borel probability invariant measure of Denjoy type arise. In fact, for any rotation number $\omega \in \mathbf{R} \setminus \mathbf{Q}$, there is a topological disk of any given dimension of Denjoy invariant measures of rotation number ω (i.e. isomorphic to the rotation R_ω), with respect to the weak topology on the space of Borel probability measures. The support of these measures is contained in the set of orbits which minimize the action *locally*, while the action minimizing sets Σ_ω corresponds to global minima. This construction has been generalized in [F], where almost periodic invariant measures, having arbitrary set of frequencies, and positive entropy invariant measures are constructed for any rotation number $\omega \in \mathbf{R} \setminus \mathbf{Q}$, in case the corresponding minimizing set Σ_ω is not an invariant curve. This is the first variational result on the dynamics in the stochastic regions which will be described below. The second result, due entirely to the first author [Ma11], consists in the variational construction of wandering orbits in a Birkhoff region of instability R. This orbits can approach infinitely many times, in the past and in the future, each Aubry-Mather set contained in the stochastic region R. We finally recall that, since the union of all minimal and locally minimal configurations in R is a closed set of zero Lebesgue measure, one cannot hope to obtain by this method positive Lyapunov exponents on a set of positive area (or equivalently positive entropy for the Lebesgue measure) in the Birkhoff regions of instability. Thus, in spite of the numerical evidences, the problem of *proving* the stochastic behaviour in these regions, in the sense of smooth ergodic theory, remains open.

Invariant measures supported within the gaps of minimizing sets.

We will outline below the proof of the following theorems:

Theorem 19.1. *Suppose that $f \in J^1$ is an exact area-preserving monotone twist diffeomorphism of the annulus $\mathbf{S}^1 \times \mathbf{R}$ which does not admit any rotational invariant curve of rotation number $\omega \in \mathbf{R} \setminus \mathbf{Q}$. Let Ω be any finite or countable set of rationally independent frequencies and let Λ_Ω be the free group generated by Ω. If Ω satisfies the condition*

$$(\Omega) \qquad\qquad \omega \in \Lambda_\Omega \ ,$$

then there exists an f-invariant ergodic almost periodic Borel probability measure μ_Ω on $S^1 \times R$, of angular rotation number ω, having Ω as set of its frequencies, i.e. the set of eigenvalues

$$\{\exp(2\pi i\omega_\ell) / \omega_\ell \in \Omega\}$$

generates the spectrum of μ_Ω. Furthermore the following localization property holds: if σ_ω is the unique f-invariant ergodic Borel probability measure supported on the Aubry-Mather set $\Sigma_\omega^\star$, then μ_Ω can be chosen to be arbitrarily close to σ_ω, with respect to the weak topology on the space of Borel probability measures with compact support on $S^1 \times R$.

Theorem 19.2. *Under the same hypotheses of Theorem 19.1, there exists an f-invariant ergodic Borel probability measure μ_h, of angular rotation number ω, having positive entropy. Furthermore the measure μ_h can be chosen arbitrarily close to the unique f-invariant ergodic measure σ_ω supported on the Aubry-Mather set $\Sigma_\omega^\star$, with respect to the weak topology on the space of compactly supported Borel probability measures on $S^1 \times R$.*

The *angular rotation number* of a f-invariant ergodic Borel probability measure μ is defined as the rotation number of almost every orbit (with respect to the measure μ) contained in the support of μ. It is well defined as a consequence of Birkhoff ergodic theorem.

Theorem 19.1 contains, as a particular case, the Denjoy type invariant measures constructed and studied by the first author in [Ma6]. They correspond to the case $\Omega = \{\omega/k\}$, $k \in Z \setminus \{0\}$. For each $k \in Z \setminus \{0\}$, there exists a $|k|$-dimensional topological disk of such Denjoy measures (with respect to the weak topology on the space of Borel probability measures), as it is shown in [Ma6]. A previous result on the entropy of twist maps was obtained by S.Angenent [An1-2], who proved that, if R is a Birkhoff region of instability, then $f|R$ has positive topological entropy. This result implies, via the variational principle for the topological entropy, the existence of positive entropy Borel probability measures whose support is contained in R. The main advantage of Theorem 19.2 consists in a finer localization of the supports of positive entropy measures.

The basic idea, contained in [Ma6], underlying the proof of Theorems 19.1 and 19.2, is to minimize the Percival's Lagrangian (6.1) subject to constraints, thereby constructing configurations which minimizes *locally* but are not minimal in general. The difficulty in realizing this plan is due to the possibility that the minimizing element does not lie in the "interior" of the constraint, so that it does not produce, as it does in §6, stationary configurations, i.e. the Euler-Lagrange equation may fail. This can be avoided when the Peierls's barrier P_ω^h is striclty positive at some point $\xi \in R$, i.e. when no rotational invariant curve of rotation number ω exists, according to Lemma 18.1.

The constraints will be given by biinfinite sequences of real numbers, i.e. by elements $\Delta \in R^Z$. Therefore the space of constraints will be a subspace of R^Z, which will be endowed with the product topology. Let in fact consider the following definition:

$$\mathcal{D}_\infty = \{\Delta \in \mathbf{R}^{\mathbf{Z}}/\|\Delta\|_\infty < \infty \text{ and } j + \Delta(j) \le j + 1 + \Delta(j+1), \text{ for all } j \in \mathbf{Z}\},$$

where $\|\cdot\|_\infty$ denotes the ℓ^∞ norm on $\mathbf{R}^{\mathbf{Z}}$. The constraints themselves will be given by the following spaces:

Definition 19.3. *Let $\mathcal{A}_\xi$ be the set of all Borel measurable functions $\phi : \mathcal{D}_\infty \times \mathbf{R} \to \mathbf{R}$ satisfying the following conditions:*

1)
$$\phi_{\sigma\Delta}(t+1) = \phi_\Delta(t) + 1 ,$$

where $\sigma : \mathcal{D}_\infty \to \mathcal{D}_\infty$ is the forward shift, i.e. $\sigma\Delta(j) = \Delta(j-1)$, for all $j \in \mathbf{Z}$.

2)
$$\begin{cases} \phi_\Delta(t) \le \xi + j, & \text{if } t \le j + \Delta(j) \\ \phi_\Delta(t) \ge \xi + j, & \text{if } t > j + \Delta(j) , \end{cases} \quad \text{for all } j \in \mathbf{Z} .$$

Definition 19.4. *Let $\mathcal{B}_\xi$ be the subset of $\mathcal{A}_\xi$ consisting of those elements which are weakly order preserving with respect to $t \in \mathbf{R}$.*

Notice that $\mathcal{D}_\infty$ is chosen to assure that Definitions 19.3 and 19.4 are well posed.

In order to better understand the meaning of the definitions just given, it is useful to consider the following fact: it is possible to *normalize* the minimizer ϕ_ω, whose existence is the content of Theorem 6.1, in such a way that it satisfies the following condition:

⋆)
$$\begin{cases} \phi_\omega(t) \le \xi + j, & \text{if } t \le j \\ \phi_\omega(t) > \xi + j, & \text{if } t > j . \end{cases}$$

This can be done since, by the translation invariance of F_ω, for any $a \in \mathbf{R}$, $\phi_\omega T_a$ minimizes F_ω over Y^*, if ϕ_ω does. Here T_a denotes the translation $t \to t+a$, $t \in \mathbf{R}$. Choosing $a = \sup\{t / \phi_\omega(t) \le \xi\}$, one obtains $\phi_\omega T_a \le \xi$, for $t \le 0$, and $\phi_\omega T_a > 0$, for $t > 0$. Since $\phi_\omega(t+1) = \phi_\omega(t) + 1$, $\phi_\omega T_a$ satisfies ⋆).

In view of ⋆) the constraints introduced in Definitions 19.3 and 19.4 can be interpreted as a sort of *noise* around the minimal configurations associated to ϕ_ω. Therefore it is natural to introduce an averaged version of the Percival's Lagrangian.

Let $\mathcal{A}^*$ be the space consisting of all Borel measurable functions $\phi : \mathcal{D}_\infty \times \mathbf{R} \to \mathbf{R}$ in $L^\infty(\mathcal{D}_\infty, L^\infty_{loc}(\mathbf{R}))$ satisfying the condition $\phi_{\sigma\Delta}(t+1) = \phi_\Delta(t) + 1$, for all $(\Delta, t) \in \mathcal{D}_\infty \times \mathbf{R}$. Let μ be a shift-invariant ergodic Borel probability measure on $\mathcal{D}_\infty$, then the *averaged Percival-Lagrange functional* is defined on $\mathcal{A}^*$ as follows:

$$F_\omega^\mu(\phi) = \int_{\mathcal{D}_\infty} \int_a^{a+1} h(\phi_\Delta(t), \phi_\Delta(t+\omega))\, dt\, d\mu(\Delta) .$$

Notice that the definition is independent of $a \in \mathbf{R}$ and therefore F_ω^μ is *translation invariant*, i.e. $F_\omega^\mu(\phi T_a) = F_\omega^\mu(\phi)$, where $T_a : (\Delta, t) \to (\Delta, t + a)$. The functional F_ω^μ coincides with the Percival-Lagrange functional considered in [Ma6], in case μ is an atomic measure. In this case, the variational problem for F_ω^μ yields in fact the Denjoy invariant measures constructed there.

Theorem 19.5. *There exists $\phi \in \mathcal{B}_\xi$, continuous from the left with respect to $t \in \mathbf{R}$, which minimizes the averaged Percival-Lagrange functional F_ω^μ over $\mathcal{A}_\xi$.*

Proof. The functional F_ω^μ is bounded from below and lower semicontinuous, since the variational principle h is bounded from below by properties (h_1) and (H_2) and Fatou lemma can be applied. Furthermore, F_ω^μ is *submodular*, i.e. it satisfies

$$(19.1) \qquad F_\omega^\mu(\phi \vee \phi') + F_\omega^\mu(\phi \wedge \phi') \leq F_\omega^\mu(\phi) + F_\omega^\mu(\phi') \ ,$$

for any ϕ, $\phi' \in \mathcal{A}_\xi$. This property is an immediate consequence of the inequality (9.1) verified by the variational principle h and it is related to the Aubry's Crossing Lemma proved in §9. Since the spaces $\mathcal{A}_\xi$ and $\mathcal{B}_\xi$, considered with the natural order induced by the real lines and the standard lattice operations $\vee$, $\wedge$, are complete lattices (i.e. each subset X has a least upper bound $\sup X$ and a greatest lower bound $\inf X$), F_ω^μ has a minimum on $\mathcal{A}_\xi$ and one on $\mathcal{B}_\xi$. Finally, the minimum on $\mathcal{B}_\xi$ is also minimum on $\mathcal{A}_\xi$, since, if ϕ^* minimizes over $\mathcal{A}_\xi$, there exists $\phi_0 \in \mathcal{B}_\xi$ such that $F_\omega^\mu(\phi_0) \leq F_\omega^\mu(\phi^*)$. This is a consequence of the submodularity and of the translation invariance of F_ω^μ. The existence of the minima depends on the following:

Lemma 19.6. *Let $F : \Lambda \to \mathbf{R}$ be a real valued functional on a complete lattice Λ. Assume F is submodular, lower semicontinuous and bounded from below. Then F has a minimum on Λ.*

Proof. Since F is bounded from below, $\ell = \inf_\Lambda F$ is a real number. Therefore, given any sequence of positive real numbers $(\epsilon_i)_{i \in \mathbf{N}}$, converging to zero, there exists a sequence $(\phi_i)_{i \in \mathbf{N}}$ of elements of Λ such that:

$$(19.2) \qquad \ell \leq F(\phi_i) \leq \ell + \epsilon_i \ .$$

By the submodularity, since $F(\phi_i \vee \phi_{i+1}) \geq \ell$,

$$F(\phi_i \wedge \phi_{i+1}) \leq \ell + \epsilon_i + \epsilon_{i+1} \ ,$$

and iterating

$$(19.3) \qquad F(\phi_i \wedge ... \wedge \phi_{i+k}) \leq \ell + \epsilon_i + ... + \epsilon_k \ .$$

Define $\psi_{i,k} = \phi_i \wedge ... \wedge \phi_{i+k}$, for all $i, k \geq 1$. Then by construction $(\psi_{i,k})_{k \in \mathbf{N}}$ is a non-increasing sequence of elements of Λ and, since Λ is complete, it has a limit $\psi_i = \inf_{k \in \mathbf{N}} \psi_{i,k}$. As a consequence of the inequality 2) and of the lower semicontinuity of F, one gets:

$$(19.4) \qquad F(\psi_i) \leq \ell + r_i \ ,$$

where $r_i = \sum_{k \geq i} \epsilon_k$ is finite and converges to zero, as $i \to +\infty$, if $(\epsilon_i)_{i \in \mathbf{N}}$ is chosen such that

$$\sum_{i \geq 0} \epsilon_i < +\infty \ .$$

On the other hand, since $\psi_{i,k} \leq \psi_{i+1,k-1}$, for all $i, k \geq 1$, then $(\psi_i)_{i \in \mathbb{N}}$ is a non-decreasing sequence, hence it has a limit $\psi \in \Lambda$. By the lower semicontinuity and the choice of the sequence $(\epsilon_i)_{i \in \mathbb{N}}$, (19.4) implies:

$$(19.5) \qquad F(\psi) \leq \ell + \liminf_{i \to +\infty} r_i = \ell \, ,$$

which concludes the proof, by showing that ψ is a minimum point for F. $\qquad \square$

Given $\phi \in \mathcal{A}^*$ and $t \in \mathbb{R}$ we define configurations $x_\Delta = x_{\phi \omega \Delta t}$ by

$$(x_\Delta)_i = \phi_\Delta(t + \omega i) \, , \quad \text{for all } i \in \mathbb{Z}$$

and, if ϕ is also weakly order preserving with respect to $t \in \mathbb{R}$, we define configurations $x_\Delta^\pm = x_{\phi \omega \Delta t \pm}$ by

$$(x_\Delta^\pm)_i = \phi_\Delta(t + \omega i \pm) \, , \quad \text{for all } i \in \mathbb{Z} \, .$$

If $\phi \in \mathcal{A}_\xi$ and $x = x_{\phi \omega \Delta t-}$, then

$$x_i \leq \xi + j \, , \quad \text{if } t + \omega i \leq j + \Delta(j) \, ,$$
$$x_i \geq \xi + j \, , \quad \text{if } t + \omega > j + \Delta(j) \, .$$

We let $\mathcal{H}_{\omega \Delta t-}$ denote the set of configurations $x \in \mathbb{R}^{\mathbb{Z}}$ which satisfy these conditions. If $\phi \in \mathcal{A}_\xi$ and $x = x_{\phi \omega \Delta t+}$, then

$$x_i \leq \xi + j \, , \quad \text{if } t + \omega i < j + \Delta(j) \, ,$$
$$x_i \geq \xi + j \, , \quad \text{if } t + \omega i \geq j + \Delta(j) \, .$$

We let $\mathcal{H}_{\omega \Delta t+}$ denote the set of configurations $x \in \mathbb{R}^{\mathbb{Z}}$ which satisfy the latter conditions. Clearly the sets of configurations just defined also depend on $\xi \in \mathbb{R}$, but we prefer to drop this dependence, since ξ will be a fixed real number to be chosen appropriately, i.e. such that $P_\omega(\xi) > 0$. Notice that, if $\phi \in \mathcal{B}_\xi$ is continuous from the left with respect to $t \in \mathbb{R}$, then

$$x_{\phi \omega \Delta t} = x_{\phi \omega \Delta t-} \in \mathcal{H}_{\omega \Delta t-} \, .$$

A configuration $x \in \mathcal{H}_{\omega \Delta t \pm}$ is said to be *minimal relative to* $\mathcal{H}_{\omega \Delta t \pm}$ if:

for any pair of integers $m < n$ and any configuration $x' \in \mathcal{H}_{\omega \Delta t \pm}$ such that $x'_m = x_m$ and $x'_n = x_n$,

$$h_{mn}(x) \leq h_{mn}(x') \, ,$$

where h_{mn} is the function defined as follows

$$h_{mn}(x) = \sum_{i=m}^{n-1} h(x_i, x_{i+1}) \, .$$

Lemma 19.7. *Let $\phi \in \mathcal{B}_\xi$ and suppose ϕ minimizes F_ω^μ over $\mathcal{B}_\xi$. Then, for any given $t \in \mathbb{R}$, there exists a full measure set $\mathcal{D}_\infty(t) \subset \mathcal{D}_\infty$, with respect to the* measure μ, such that the configuration $x_{\phi\omega\Delta t\pm}$ is minimal relative to $\mathcal{H}_{\omega\Delta t\pm}$, for all $\Delta \in \mathcal{D}_\infty(t)$.

Thus a minimizing element $\phi \in \mathcal{B}_\xi$ for the averaged Percival's Lagrangian can be used to produce an abundance of relatively minimal configurations. Assuming that $P_\omega(\xi) > 0$, it is plausible that the minimizer ϕ can be chosen satisfying *strictly* the inequalities defining $\mathcal{A}_\xi$. In fact, if it were not so, we could produce relatively minimal configurations $x = x_{\phi\omega\Delta t\pm}$, satisfying $x_0 = \xi + j$, which is impossible since $P_\omega(\xi) > 0$. This fact is made precise in the following:

Lemma 19.8. *Suppose the generating function h satisfies the conditions (H_1) – $(H_{6\theta})$. If ω is irrational, $P_\omega(\xi) > 0$ and μ is a shift-invariant ergodic Borel probability measure supported on $\mathcal{D}_c$, $0 < c < \delta_\theta(\omega, P_\omega(\xi))$, where $\mathcal{D}_c = \{\Delta \in \mathcal{D}_\infty \mid \|\Delta\|_\infty \leq c\}$, then there exists $\phi \in \mathcal{B}_\xi$ minimizing F_ω^μ over $\mathcal{A}_\xi$, which satisfies strictly the inequalities defining $\mathcal{A}_\xi$, i.e.*

$$2') \qquad \begin{cases} \phi_\Delta(t) < \xi + j\,, & \text{if } t \leq j + \Delta(j) \\ \phi_\Delta(t) > \xi + j\,, & \text{if } t > j + \Delta(j)\,, \end{cases} \quad \text{for all } j \in \mathbf{Z}\,,$$

for almost all $(\Delta, t) \in \mathcal{D}_\infty \times \mathbf{R}$, with respect to the measure $\mu_\mathcal{L} = \mu \times \text{Lebesgue}$.

In view of Lemma 19.8, the proof of the Euler-Lagrange equation is immediate and it does not deviate from the usual argument.

Lemma 19.9. *Suppose the same conditions as in Lemma 19.8 hold. Then ϕ satisfies the Euler-Lagrange equation, i.e.*

$$h_2(\phi_\Delta(t - \omega), \phi_\Delta(t)) + h_1(\phi_\Delta(t), \phi_\Delta(t + \omega)) = 0\,,$$

for almost all $(\Delta, t) \in \mathcal{D}_c \times \mathbf{R}$, with respect to $\mu_\mathcal{L} = \mu \times \text{Lebesgue}$.

Let f be an exact area-preserving monotone twist map of the annulus $\mathbf{S}^1 \times \mathbf{R}$. Let h be a variational principle associated to f. The conditions imposed on f imply that h satisfies the conditions (H_1) – $(H_{6\theta})$, for some $\theta > 0$. Let $\omega \in \mathbf{R} \setminus \mathbf{Q}$ and assume that there is no rotational f-invariant curve of rotation number ω, hence $P_\omega(\xi) > 0$, for some $\xi \in \mathbf{R}$, by Lemma 18.1, and consequently Lemma 19.8 and 19.9 hold.

Let $0 < c < \delta_\theta(\omega, P_\omega(\xi))$ and let μ be a shift-invariant ergodic Borel probability measure on $\mathcal{D}_c$. We will introduce a dynamical system $(\mathcal{R}, \mu_\mathcal{L})$, which is the suspension of (σ, μ) over the rotation R_ω, where σ is, as before, the forward shift on $\mathbf{R}^\mathbf{Z}$ and R_ω is the rigid rotation $t \to t + \omega$ (mod. $\mathbf{Z}$).

Let $\mathcal{S}_c$ be the quotient space $\mathcal{S}_c = (\mathcal{D}_c \times \mathbf{R})/T$, where $T : \mathcal{D}_c \times \mathbf{R} \to \mathcal{D}_c \times \mathbf{R}$ is the transformation $T : (\Delta, t) \to (\sigma\Delta, t + 1)$. The measure $\mu_\mathcal{L}$ induces on $\mathcal{S}_c$ a Borel probability measure, still denoted by the same symbol. The Euler-Lagrange equation suggests to consider on the measure space $(\mathcal{S}_c, \mu_\mathcal{L})$ the transformation $\mathcal{R} : (\Delta, t) \to (\Delta, t + \omega)$ (mod. T), induced by R_ω, which preserves the measure $\mu_\mathcal{L}$. We formalize this construction in the following:

Definition 19.10. *If (X, T, μ) is a dynamical system on a measure space $(X, \mathcal{B})$, leaving the probability measure μ invariant, its suspension over the rotation R_ω is the system $(\mathcal{S}, \mathcal{R}, \mu_\mathcal{L})$ defined as follows:*

1) $\mathcal{S} = (X \times \mathbf{R})/T$, where $T(\tilde{x}, t) = (\mathcal{T}x, t+1)$, for any $(x, t) \in X \times \mathbf{R}$ and the measure $\mu_\mathcal{L}$ is the probability measure induced by the product measure $\mu \times \mathcal{L}$, where $\mathcal{L}$ is the Lebesgue measure on $\mathbf{R}$;

2) $\mathcal{R}$ is the transformation induced on $\mathcal{S}$ by the rotation R_ω, i.e. $\mathcal{R} : (x, t) \to (x, t + \omega) \pmod{.T}$.

The minimizing element $\phi \in \mathcal{B}_\xi$ yielded by the previous arguments induces a Borel map $\Phi : \mathcal{S}_c \to \mathbf{S}^1 \times \mathbf{R}$ which semi-conjugates f to $\mathcal{R}$ on $\mathcal{S}_c$. In fact, let

$$\eta_\Delta = -h_1(\phi_\Delta(t), \phi_\Delta(t+\omega)) = h_2(\phi_\Delta(t-\omega), \phi_\Delta(t))$$

and

$$\Phi : (\Delta, t) \to (\phi_\Delta(t), \eta_\Delta(t)) \pmod{. \mathbf{Z} \times \{\mathrm{id}\}} ,$$

then

$$f \circ \Phi = \Phi \circ \mathcal{R} ,$$

almost everywhere on $\mathcal{S}_c$, with respect to $\mu_\mathcal{L}$. Therefore the measure $\mu_\mathcal{L}$ induces by push-forward an f-invariant Borel probability measure μ_ω on $\mathbf{S}^1 \times \mathbf{R}$, i.e. $\mu_\omega = \Phi_*(\mu_\mathcal{L})$. The condition for the measure $\mu_\mathcal{L}$ to be ergodic, with respect to the transformation $\mathcal{R}$, and consequently for the measure μ_ω to be ergodic with respect to the diffeomorphism f (since the push-forward of an ergodic measure is still ergodic), is the following:

$$(E_\omega) \qquad \exp(2\pi i k/\omega) \notin EV(\mu) , \quad \text{for all } k \in \mathbf{Z} \setminus \{0\} ,$$

where $EV(\mu)$ is the eigenvalues spectrum of (σ, μ). This can be seen by a spectral theory argument which shows that the only eigenfunctions corresponding to the eigenvalue 1 are the constant functions.

The previous construction can be summarized as follows:

Theorem 19.11. *Let $\omega \in \mathbf{R} \setminus \mathbf{Q}$ and suppose that the diffeomorphism f does not admit any rotational invariant curve of rotation number ω. Then any shift-invariant ergodic Borel probability measure μ on $\mathcal{D}_c$, $0 < c < \delta_\theta(\omega, P_\omega(\xi))$, satisfying the condition (E_ω), induces an f-invariant ergodic Borel probability measure μ_ω on $\mathbf{S}^1 \times \mathbf{R}$, of angular rotation number ω. The resulting dynamical system (f, μ_ω) can be described as a factor of the suspension $(\mathcal{R}, \mu_\mathcal{L})$ of (σ, μ) over the rigid rotation R_ω. The factor map $\Phi : \mathcal{S}_c \to \mathbf{S}^1 \times \mathbf{R}$ is described as:*

$$\Phi(\Delta, t) = (\phi_\Delta(t), \eta_\Delta(t)) \pmod{. \mathbf{Z} \times \{\mathrm{id}\}} ,$$

where $\phi : \mathcal{D}_c \times \mathbf{R} \to \mathbf{R}$ is weakly order preserving, with respect to $t \in \mathbf{R}$, it satisfies the following localization property:

$$\phi_\omega(t - c) \le \phi_\Delta(t) \le \phi_\omega(t + c) ,$$

and the $\mathcal{A}_\xi$ constraints strictly:

1)
$$\phi_{\sigma\Delta}(t+1) = \phi_\Delta(t) + 1 \ ,$$

and

2')
$$\begin{cases} \phi_\Delta(t) < \xi + j, & \text{if } t \le j + \Delta(j) \\ \phi_\Delta(t) > \xi + j, & \text{if } t > j + \Delta(j) \ , \quad \text{for all } j \in \mathbf{Z} \ , \end{cases}$$

for almost all $(\Delta, t) \in \mathcal{D}_c \times \mathbf{R}$, with respect to the measure $\mu_{\mathcal{L}}$.

As before, σ denotes the forward shift on $\mathbf{R}^{\mathbf{Z}}$ and $\phi_\omega \in Y_1$ minimizes F_ω according to Theorem 6.1 and it is normalized as in $\star$). Finally $\xi \in \mathbf{R}$ is chosen such that $P_\omega(\xi) > 0$.

According to Theorem 19.11, it is possible to associate to any shift-invariant Borel probability measure μ on $\mathcal{D}_c$ a f-invariant Borel probability measure μ_ω on $\mathbf{S}^1 \times \mathbf{R}$. The question naturally raised by the previous construction concerns the ergodic-theoretical properties or classification of the measures μ_ω, which can be obtained through it. Since μ_ω is given as a *factor*, it is not straigthforward to understand its nature in general. However, the factor map is partially controlled by the information provided by the constraints. This is enough to conclude that if μ is an almost periodic measure, satisfying condition (E_ω), then the factor map is an isomorphism, thereby obtaining Theorem 19.1. This case includes the case when μ is an atomic shift-invariant probability measure, which give the Denjoy f-invariant probability measures constructed in [Ma6]. Furthermore, it is possible to choose (σ, μ) isomorphic to a Bernoulli shift on two symbol, in such a way that the entropy of the associated f-invariant measure μ_ω is also positive. This is the content of Theorem 19.2. The details of these arguments are contained in [F]. We finally remark that the freedom in the choice of $c > 0$ in Theorem 19.11 can be used to localize the measure μ_ω constructed there, as claimed in Theorems 19.1 and 19.2.

We will briefly sketch the basic idea underlying the construction of positive entropy measures. According to Lemma 19.8 and 19.9, there exists $\phi : \mathcal{D}_c \times \mathbf{R} \to \mathbf{R}$ which satisfies the constraints strictly and therefore the Euler-Lagrange equation, if $c > 0$ is chosen sufficiently small. Let A_- and A_+ be the subset of the cylinder $\mathbf{S}^1 \times \mathbf{R}$ defined as follows:

$$A_- = \{(\theta, y) | \xi - 1/2 \le \theta < \xi \,(\text{mod}.\mathbf{Z})\} \ , \ A_+ = \{(\theta, y) | \xi < \theta \le \xi + 1/2 \,(\text{mod}.\mathbf{Z})\} \ .$$

Consider relatively minimal configurations $x_\Delta^\pm$ defined by $(x_\Delta^\pm)_i = \phi_\Delta(t + i\omega\pm)$ and the corresponding orbits $\left((x_\Delta^\pm)_i, (y_\Delta^\pm)_i\right)$, $i \in \mathbf{Z}$, for the lift of f to the universal cover. If $r \in \mathbf{Z}$ is such that $t + r\omega \in (-c, c) \,(\text{mod. } \mathbf{Z})$, i.e. there exists $j_r \in \mathbf{Z}$ such that $t + r\omega - j_r \in (-c, c)$, then it possible to choose $\Delta(j_r)$ such that $|\Delta(j_r)| < c$ and either $t + r\omega < j_r + \Delta(j_r)$ or $t + r\omega > j_r + \Delta(j_r)$ is achieved. In the first case, since ϕ satisfies the constraints, $\left((x_\Delta^\pm)_r, (y_\Delta^\pm)_r\right)$ belongs to the lift of A_- to the universal cover. In the second case it belongs to the lift of A_+. Thus, we are able to construct, by appropriate choices of the constraint $\Delta \in \mathcal{D}_c$, orbits for the diffeomorphism f which belongs *arbitrarily* to A_- or A_+ each time the orbit $R_\omega^i(t)$,

$i \in \mathbf{Z}$, of the rigid rotation $R_\omega : t \to t + \omega$ (mod. $\mathbf{Z}$) belongs to the interval $(-c, c)$. By the irrationality of ω, R_ω is ergodic with respect to the Lebesgue measure. Thus, $R_\omega^i(t) \in (-c, c)$ with frequency equal to $2c$. Therefore the entropy of the map f is positive and proportional to c.

Chaotic orbits in a Birkhoff region of instability.

It is a consequence of Birkhoff invariant curve theorem that, in a Birkhoff region of instability R, for an exact area-preserving monotone twist diffeomorphism of the annulus f, there are orbits which connect two preassigned open neighborhoods of the two connected components of ∂R. This fact has already been noticed in the proof of the existence, for any $\epsilon > 0$, of ϵ-glancing trajectories of convex plane billiards in a domain whose boundary has zero curvature at some point. We are refering to Lemma 17.2. This statement can be significantly strengthened by a variational construction of orbits based on the positivity of the Peierls's barriers $P_\omega^h(\xi)$, corresponding to rotation symbols $\omega \in (\rho(\Gamma_-), \rho(\Gamma_+))$, where $\Gamma_\pm$ are the two rotational invariant curves giving the boundary of the Birkhoff region of instability. The main idea consists, as in the previous construction of invariant measures, in minimizing the "energy" associated to the variational principle h over configurations subject to constraints. The positivity of all Peierls's barriers in the region R will assure that, for a wide but appropriate choice of constraints, the resulting minimizing configurations will be contained in the interior of the constraint, thereby satifying the stationarity condition. We recall that stationary configurations are in one-to-one correspondence with orbits of the diffeomorphism f associated to the variational principle h. The results which we will describe are contained in [Ma11]. We recall that a *Birkhoff region of instability* is a compact f-invariant subset of the infinite cylinder, satisfying the following properties: (1) ∂R consists of two connected components Γ_- and Γ_+, which are rotational f-invariant curves; (2) if Γ is a rotational invariant curve contained in R, then $\Gamma = \Gamma_-$ or $\Gamma = \Gamma_+$.

Theorem 19.12. *If $\rho(\Gamma_-) \le \omega_-, \omega_+ \le \rho(\Gamma_+)$, then there is an f-orbit $\mathcal{O}$ in R such that $\mathcal{O}$ is α-asymptotic (i.e. asymptotic in the past) to $\Sigma_{\omega_-}^\star$ and ω-asymptotic (i.e. asymptotic in the future) to $\Sigma_{\omega_+}^\star$, provided that if ω_- (resp. ω_+) $= \rho(\Gamma_-)$ (resp. $\rho(\Gamma_+)$), then ω_- (resp. ω_+) is irrational. Here $\Sigma_\omega^\star$ denotes, as before, the Aubry-Mather set of rotation symbol ω, described in §14.*

The statement that $\mathcal{O} = \{(\theta_i, y_i)\}_{i \in \mathbf{Z}}$ is α-asymptotic in R to $\Sigma_{\omega_-}^\star$ means that dist$((\theta_i, y_i), \Sigma_{\omega_-}^\star) \to 0$ as $i \to -\infty$. The statement that $\mathcal{O}$ is ω-asymptotic to $\Sigma_{\omega_+}^\star$ means that dist$((\theta_i, y_i), \Sigma_{\omega_+}^\star) \to 0$ as $i \to +\infty$.

Theorem 19.13. *Consider for each $i \in \mathbf{Z}$, a real number $\rho(\Gamma_-) \le \omega_i \le \rho(\Gamma_+)$ and a positive number ϵ_i. There exists an f-orbit $\mathcal{O} = \{(\theta_j, y_j)\}_{j \in \mathbf{Z}}$ in R and an increasing bi-infinite sequence $..., j(i), ...$ of integers such that dist$((\theta_{j(i)}, y_{j(i)}), \Sigma_{\omega_i}^\star) < \epsilon_i$.*

In other words, $\mathcal{O}$ approaches within $\epsilon_i > 0$ of $\Sigma_{\omega_i}^\star$ at the $j(i)^{th}$ iteration.

By a *constraint* $\mathcal{J}$, we will take in this case a bi-infinite sequence $(..., J_i, ...)$, where each J_i is a closed, connected, non-empty subset of $\mathbf{R}$. A $\mathcal{J}$-*configuration* will be a biinfinite sequence $(x_i)_{i \in \mathbf{Z}}$, with $x_i \in J_i$. A *segment of a $\mathcal{J}$-configuration* will be a finite sequence $(x_j, ..., x_k)$ such that $x_i \in J_i$, for each $j \le i \le k$. Let h be a variational principle associated to an exact area-preserving monotone twist map f. A segment $(x_j, ..., x_k)$ of a $\mathcal{J}$-configuration will be said to be $\mathcal{J}$-*minimal* (with respect to h) if

$$h(x_j, ..., x_k) \leq h(x_j^*, ..., x_k^*) ,$$

for every segment of a $\mathcal{J}$-configuration $(x_j^*, ..., x_k^*)$ such that $x_j^* = x_j$ and $x_k^* = x_k$. We will say that a $\mathcal{J}$-configuration $(x_i)_{i \in \mathbf{Z}}$ is $\mathcal{J}$-minimal if for every $j < k$, the corresponding segment $(x_j, ..., x_k)$ of it is $\mathcal{J}$-minimal.

It is not difficult to specify conditions on the constraint $\mathcal{J}$ which assures the existence of $\mathcal{J}$-minimal configurations.

Lemma 19.14. *Let $(..., J_i, ...)$ be a constraint such that there exist arbitrarily small and arbitrarily large i for which J_i is bounded. Then there exists a $\mathcal{J}$-minimal configuration.*

Proof. By properties (H_1) and (H_2), the function $h(x_{-N}, x_{-N+1}, ..., x_N)$ is proper, continuous and bounded below on $J_{-N} \times J_{-N+1} \times ... \times J_N$. It follows that there is a sequence $(x_{-N}^{(N)}, x_{-N+1}^{(N)}, ..., x_N^{(N)})$ which minimizes this function over $J_{-N} \times J_{-N+1} \times ... \times J_N$. Furthermore, using (H_2), one may find, for each integer j, a compact set K_j such that $x_j^{(N)} \in K_j$ for all N: if J_j is bounded one takes $K_j = J_j$. Otherwise, it follows from the fact that there exist $j' < j < j''$ for wich $J_{j'}$ and $J_{j''}$ are bounded and from property (H_2) of the variational principle h. By a compactness argument combined with the Cantor diagonal process, it is possible to choose a sequence $(N_i)_{i \in \mathbf{N}}$ such that the subsequence $x_j^{(N_i)} \to x_j \in K_j$, as $i \to +\infty$. It is straigthforward to verify that the limiting configuration $(..., x_j, ...)$ is a $\mathcal{J}$-minimal configuration. $\qquad\square$

$\mathcal{J}$-minimal configurations are not always stationary, i.e. they do not yield always orbits of the diffeomorphism f. This will happen in case the $\mathcal{J}$-configuration is also $\mathcal{J}$-*free*, i.e. $x_i \in \text{int } J_i$, for each i.

Lemma 19.15. *Let $x = (..., x_i, ...)$ be a $\mathcal{J}$-minimal configuration. If x is $\mathcal{J}$-free, then it is stationary, i.e. $-\partial_1 h(x_i, x_{i+1}) = \partial_2 h(x_{i-1}, x_i)$. In particular, $f(x_i, y_i) = (x_{i+1}, y_{i+1})$, where $y_i = -\partial_1 h(x_i, x_{i+1})$, i.e. $(x_i, y_i)_{i \in \mathbf{Z}}$ is an f-orbit.*

Proof. Since x is $\mathcal{J}$-free, any sufficiently small variation of the configuration x will still be a $\mathcal{J}$-configuration. Thus, the $\mathcal{J}$-minimality of x implies its stationarity by a simple standard argument. $\qquad\square$

The orbits whose existence is asserted in Theorems 19.12 and 19.13 will be constructed as the orbits associated to $\mathcal{J}$-minimal and $\mathcal{J}$-free configurations. The method of proof consists in using the positivity of the Peierls's barriers, for any rotation symbol in a given range, in such a way that, if $\mathcal{J}$ is a properly chosen constraint, each $\mathcal{J}$-minimal configuration will be $\mathcal{J}$-free, and furthermore the corresponding orbits have the properties required in Theorems 19.13 and 19.14. The specifications on $\mathcal{J}$ which produce at least *partially* $\mathcal{J}$-free configurations, i.e. $\mathcal{J}$-free on some subinterval of $\mathbf{Z}$, are quite complicated. The following holds. Let ω be a real number and let $\phi_\omega \in Y_1$ be a function which minimizes the Percival's Lgrangian, according to Theorem 6.1. Let x be a minimal configuration of rotation number ω, which corresponds to a *recurrent* orbit of f, i.e. $x_i = \phi_\omega(t + \omega i \pm)$ for some $t \in \mathbf{R}$. We also choose a real number a such that $P^h_\omega(a) > 0$, which is possible by Lemma 18.1, since there is no rotational invariant curve of rotation number ω. For each integer i, we let a_i be the unique real number such that $a_i - a \in \mathbf{Z}$ and $x_i \in (a_i, a_i + 1)$. This is possible, since $x_i - a$ is never an integer because $P^h_\omega(a) > 0$ and the Peierls's barrier is a periodic function. Furthermore, we will associate to certain pairs (ω, a) of real numbers an integer $K(\omega, a)$.

Lemma 19.16. *Let $\mathcal{J} = (..., J_i, ...)$ be a constraint. Let $j_0 \leq j_1$ be integers, and let ω, a be real numbers. Suppose that $K(\omega, a)$ is defined and $J_i = [a_i, a_i + 1]$ for $j_0 - K(\omega, a) \leq i \leq j_1 + K(\omega, a)$, with a_i defined as above. Let $\xi = (\xi_i)_{i \in \mathbf{Z}}$ be a $\mathcal{J}$-minimal configuration. Then*

$$a_i < \xi_i < a_i + 1 , \quad \text{for } j_0 \leq i \leq j_1 .$$

This statement will represent the typical situation, i.e. we define a constraint $\mathcal{J}$ over a certain set of integers in term of a recurrent minimal configuration (in this case x) and we prove that a $\mathcal{J}$-minimal configuration is partially $\mathcal{J}$-free, meaning that it is $\mathcal{J}$-free on a smaller range of integers.

The integer $K(\omega, a)$ is defined as follows. If $P^h_\omega(a) > 0$, we let k be the smallest integer $k > 2\theta/P^h_\omega(a)$. Let n be the smallest integer such that $k < q_n$, where $(p_n/q_n)_{n \in \mathbf{N}}$ are the convergents of the continued fraction expansion of the real number ω. If such integer n exists (i.e. if ω is irrational or $\omega = p/q$, in lowest terms, with $k < q$), we set $K(\omega, a) = q_{n-1} + q_n$ and it will be undefined otherwise. We give below the proof of Lemma 19.16.

Proof. The argument can be reduced to the case $j_0 = j_1$, without restriction. We write j for the common value $j_0 = j_1$. Furthermore we will prove only that $\xi_j < a_j + 1$, since the other inequality $\xi_j > a_j$ can be obtained in a similar way. Let s be the unique real number such that $\phi_\omega(s + \omega j -) < a_j + 1 < \phi_\omega(s + \omega j +)$. The real number s exists since $P^h_\omega(a_j + 1) = P^h_\omega(a) > 0$, ϕ_ω is order preserving and satisfies $\phi_\omega(s+1) = \phi_\omega(s) + 1$. Let $y_i = \phi_\omega(s + \omega i -)$. Thus y defines a minimal configuration of rotation number ω. Furthermore, since $x_i = \phi_\omega(t + \omega i \pm)$ and $x_i \in (a_i, a_i + 1)$, clearly $t \leq s$ and $x \leq y$, as configurations.

The first step in the proof consists in the case when there exist i_0 satisfying $j - K \leq i_0 < j$ and i_1 satisfying $j < i_1 \leq j + K$ such that $a_i + 1 < y_i$ for $i = i_0, i_1$. Here we denoted by K the real number $K(\omega, a)$. In this case $\xi_i \leq a_i + 1 < y_i$, for $i = i_0, i_1$, and it follows from an adapted version of Aubry's Crossing Lemma, which states that the Aubry graphs of minimal configurations cross at most once, that $\xi_i < y_i$, for $i_0 \le i \le i_1$, and in particular $\xi_j < y_j < a_j + 1$, which is what it was to be proved. The difference with the standard Aubry's Crossing Lemma established in §9 consists in the fact that the configurations y and ξ are not both minimal, since ξ is in fact $\mathcal{J}$-minimal, i.e. minimal among configurations subject to the $\mathcal{J}$-constraint. However, it is not difficult to adapt the proof given in §9 to deal with this slightly more general case.

In the general situation, we can assume $y_i < a_i + 1$ for $j \le i \le j + K$ or for $j - K \le i \le j$. In fact, if it is not so we can apply the previous argument. It is not restrictive to assume that the first alternative holds, since the other is similar. We set

$$(19.6) \qquad z_i = \phi_\omega(s + \|q_{n-1}\omega\| + \omega i+) \,,$$

where, as before, p_n/q_n denotes the nth approximant of the continued fraction expansion of the real number ω and $\|\lambda\|$ denotes in this context the distance of the real number λ from the closest integer. By its definition, $z = (z_i)_{i \in \mathbf{N}}$ is a minimal configuration. Since, by the standard properties of the continued fraction expansion, q_n is defined as the smallest integer $q > 0$ such that $\|q\omega\| < \|q_{n-1}\omega\|$, the projections in $\mathbf{R}/\mathbf{Z}$ of the intervals $[(i - j)\omega, (i - j)\omega + \|q_{n-1}\omega\|]$, $i = j, ..., j + q_n - 1$ do not overlap. Since ϕ_ω is injective and satisfies $\phi_\omega(s + 1) = \phi_\omega(s) + 1$, this implies that the projections in $\mathbf{R}/\mathbf{Z}$ of the intervals $[y_i, z_i]$, $i = j, ..., j + q_n - 1$, do not overlap. Let k be the smallest integer $> 2\theta/P_\omega^h(a)$ in the definition of $K(\omega, a)$. Since q_n is chosen so that $k \le q_n - 1$, it follows that there exists i_1 satisfying $j < i_1 \le j + k$ such that $z_{i_1} - y_{i_1} \le k^{-1}$. Since the projection in $\mathbf{R}/\mathbf{Z}$ of $[y_i, z_i]$ does not overlap that of $[y_j, z_j]$, for $j < i \le i_1$, and $y_j < a_j + 1 < z_j$, by definition of y_i and z_i, it follows that $a_i + 1$ does not belong to $[y_i, z_i]$. Since, as we assumed, $y_i < a_i + 1$, we then obtain $z_i < a_i + 1$, for $j < i \le i_1$.

Let i_2 be the smallest value of $i > j$ such that the interval $[(i - j)\omega, (i - j)\omega + \|q_{n-1}\omega\|]$ contains an integer. We have $i_1 < i_2 \le j + K$, in fact , $i_2 = j + q_n$ or $i_2 = j + q_n + q_{n-1}$, as a consequence of the fact that q_n is defined as the smallest integer $q > 0$ such that $\|q\omega\| < \|q_{n-1}\omega\|$. Therefore z is a minimal configuration satisfying $a_j + 1 < z_j$, $a_{i_2} + 1 < z_{i_2}$ and $a_i < z_i < a_i + 1$, for $j < i < i_2$. This is because $a_i \equiv a \pmod{\mathbf{Z}}$, for all $i \in \mathbf{Z}$, by definition of the configurations y and z, and by the assumption that $y_i < a_i + 1$, for $j \le i \le j + K$.

Since ξ is $\mathcal{J}$-minimal , we obtain (using as before the adapted version of the Aubry's Crossing Lemma) that $\xi_i < z_i$ for $j \le i \le i_2$. In particular $\xi_{i_1} < z_{i_1} < y_{i_1} + k^{-1}$. To summarize, we have obtained the following.

There exists $j < i_1 \le j + K$ such that

$$(19.7) \qquad \xi_{i_1} < y_{i_1} + k^{-1} \quad \text{and} \quad y_i < a_i + 1 \,, \quad \text{for } j \le i < i_1 \,.$$

Similarly, there exists $j - K \le i_0 < j$ such that

$$(19.7') \qquad \xi_{i_0} < y_{i_0} + k^{-1} \quad \text{and} \quad y_i < a_i + 1 \,, \quad \text{for } i_0 < i \le j \,.$$

(If there exists $j - K \le i_0 < j$ such that $a_{i_0} + 1 < y_{i_0}$, then clearly $\xi_{i_0} < y_{i_0}$; otherwise the argument we have just given applies).

We will argue by contradiction. We set $w_i = \xi_i$, $i = i_0$, i_1; $w_i = \min(y_i, \xi_i)$, $i_0 < i < i_1$. Assuming $\xi_j = a_j + 1$, we will obtain $h(w_{i_0}, ..., w_{i_1}) < h(\xi_{i_0}, ..., \xi_{i_1})$, contradicting the $\mathcal{J}$-minimality of the configuration ξ. We also introduce auxiliary configurations by $v_i = y_i$, for $i = i_0$, i_1, $v_i = \max(y_i, \xi_i)$, $i_0 < i < i_1$, and $\tilde{w} = y \wedge \xi$, $\tilde{v} = y \vee \xi$. Note that v_i and $\tilde{v}_i$ are the same, except at the endpoints $i = i_0$ and $i = i_1$, where they differ by an error of at most k^{-1}. The same holds for $\tilde{w}_i$ and w_i. Using the properties (H_θ) of the variational principle h, it is not difficult to prove that

$$(19.8) \qquad h(w) - h(\tilde{w}) + h(v) - h(\tilde{v}) \le 2\theta k^{-1} .$$

Thus, we replaced the segment of configurations w and v, which have the same endpoints as resp. ξ and y, by the segments $\tilde{w}$ and $\tilde{v}$ for which the formula (9.1) (Aubry's Crossing Lemma) holds:

$$(19.9) \qquad h(\tilde{w}) + h(\tilde{v}) \le h(\xi) + h(y) .$$

Thus, $h(w) \le h(\tilde{w}) + h(\tilde{v}) - h(v) + 2\theta k^{-1} \le h(\xi) + h(y) - h(v) + 2\theta k^{-1}$, where the first inequality is a consequence of (19.8) and the second follows from (19.9). Furthermore, the assumption $\xi_j = a_j + 1$ clearly implies $v_j = a_j + 1$, and we also have $v_i = y_i$, for $i = i_0$, i_1. Since y is a minimal configuration of rotation symbol ω defined by $y_i = \phi_\omega(s + \omega i -)$, where $\phi_\omega(s + \omega j -) < a_j + 1 < \phi_\omega(s + \omega j +)$, it follows that $h(v) - h(y) \ge P^h_\omega(a_j + 1) = P^h_\omega(a) > 2\theta k^{-1}$, by the choice of the integer k. Therefore,

$$(19.10) \qquad h(w) \le h(\xi) - P^h_\omega(a) + 2\theta k^{-1} < h(\xi) ,$$

which gives the announced contradiction with the $\mathcal{J}$-minimality of the configuration ξ. $\qquad\square$

The proof of Lemma 19.16 illustrates the use of the Peierls's barrier in proving that configurations which minimize, subject to appropriate constraints, are in fact locally minimal. In particular, a similar argument yields the proof of Lemma 19.8. This method is based on the fact that minimizing configurations (subject to constraint), which are localized nearby minimal configurations of rotation symbol ω, tend to avoid the regions where the Peierls's barrier $P_\omega(\xi)$ is sufficiently positive. It is then sufficient to choose constraints in such a way that they produce minimizing configurations localized nearby a minimal configuration and whose boundary lie in the region where the Peierls's barrier is sufficiently positive.

In the choice of the appropriate constraints for which Lemma 19.16 holds, a very important role is played by the choice of a recurrent minimal configuration x of rotation number ω and by a "barrier" $a \in \mathbf{R}$ such that $P^h_\omega(a) > 0$. In fact, the constraint $\mathcal{J}$ is completely specified in the range $j_0 - K \le i \le j_1 + K$ by the configuration x and the number a. One may say that the J_i's "follow" the recurrent configuration x in that range. Nevertheless, there are other recurrent minimal configurations y that the intervals J_i also follow. There is, in fact, an open interval Ω such that, for any $\omega \in \Omega$, there is a recurrent configuration y of rotation number ω such that the J_i's follow the configuration y in the range

$j_0 - K \leq i \leq j_1 + K$. Thus it is possible to have constraints which follow recurrent minimal configurations with different rotation numbers in overlapping subintervals of $\mathbf{Z}$. The definition of constraints $\mathcal{J}$ depend on the choice of a "barrier" $a \in \mathbf{R}$ and of a bi-infinite sequence of integers $(n_i)_{i \in \mathbf{Z}}$, by setting $a_i = a + n_i$. We would like to specify conditions on the sequence $(..., n_i, ...)$ which assures that, by appropriately chosing a so that $P_\omega(a) > 0$, for any $\omega \in \Omega$, then any $\mathcal{J}$-minimal configuration is $\mathcal{J}$-free.

Let $\epsilon > 0$. A bi-infinite sequence of integers $(n_i)_{i \in \mathbf{Z}}$ is said to be ϵ-restrained, if for each $j \in \mathbf{Z}$, there exist real numbers ω_j and s_j such that the following holds. Let $(p_{jn}/q_{jn})_{n \in \mathbf{N}}$ be the sequence of convergents of the continued fraction expansion of the real number ω_j. We suppose that ω is irrational or it is rational with denominator (in lowest terms) $> \epsilon^{-1}$. Under this conditions, it is possible to introduce the smallest integer $\ell(j)$ such that $q_{j,\ell(j)} > \epsilon^{-1}$. We let $K_j = q_{j,\ell(j)-1} + q_{j,\ell(j)}$. We require that the sequence $(..., n_i, ...)$ satisfies $n_i < \omega_j i + s_j < n_i + 1$, for $j - K_j \leq i \leq j + K_j$. If, furthermore, $\omega_j \in \Omega$, where Ω is some open interval, then $(n_i)_{i \in \mathbf{Z}}$ is said to be (ϵ, Ω)-restrained.

Clearly, if $(..., n_i, ...)$ is ϵ-restrained and $a \in \mathbf{R}$, the constraint $\mathcal{J} = (..., J_i, ...)$, given as before by $J_i = (a_i, a_i + 1)$, where $a_i = a + n_i$ follows a recurrent minimal configuration $x^{(j)}$ of rotation number ω_j in the interval $j - K_j \leq i \leq j + K_j$, for any $j \in \mathbf{Z}$. These configurations are given by $x_i^{(j)} = \phi_{\omega_j}(s_j + i\omega_j\pm)$, where ϕ_{ω_j} is normalized by the appropriate translation to satisfy $\phi_{\omega_j}(0-) < a < \phi_{\omega_j}(0+)$, so that $x_i^{(j)} \in (a_i, a_{i+1})$. It is then a straigthforward application of Lemma 19.16 to obtain the following:

Lemma 19.17. *Let* $\mathbf{n} = (..., n_i, ...) \in \mathbf{Z}^{\mathbf{Z}}$. *Let* $\epsilon = k^{-1}$, *where* k *is a positive integer. Let* Ω *be an open interval. Let* $a \in \mathbf{R}$ *be chosen such that*

$$P_\omega^h(a) > 2\theta\, k^{-1} ,$$

for all $\omega \in \Omega$. *Suppose that* $\mathbf{n}$ *is* (ϵ, Ω)-restrained. *Let* $a_i = a + n_i$, $J_i = [a_i, a_{i+1}]$ *and* $\mathcal{J} = (..., J_i, ...)$. *Then every* $\mathcal{J}$-minimal configuration is $\mathcal{J}$-free and therefore it satisfies the stationarity condition and corresponds to an orbit of the diffeomorphism f.

Thus, the problem of finding appropriate constraints is reduced, in view of Lemma 19.17, to the purely number theoretical problem of finding ϵ-restrained bi-infinite sequences of integers. If M is a real number, we let $F(M)$ be the set of rational numbers whose denominator (in lowest terms) is less or equal to M. We will call, following [Ma11], a connected component of the complement of $F(M)$ an *open Farey interval of heigth* M, in analogy with Farey series of number theory. If $n = (..., n_i, ...)$ is ϵ-restrained, an open Farey interval $(p/q, p'/q')$ of heigth ϵ^{-1} is said to be the *Farey interval associated to* n if the intervals

(19.11) $$A_n[j, k] = ((n_k - n_j - 1)/(k - j), (n_k - n_j + 1)/(k - j))$$

intersects $(p/q, p'/q')$ for all $j < k$. This definition are motivated by the following criterion. Let

(19.12)
$$B_n[j, k] = \bigcap_{j \leq j' < k' \leq k} A_n[j', k'] .$$

Lemma 19.18. If $\omega \in B_n[j, k]$, there exists $s \in \mathbf{R}$ such that

$$n_i < \omega i + s < n_i + 1 , \quad \text{for } j \leq i \leq k .$$

Lemma 19.18'. Let $n \in \mathbf{Z}^{\mathbf{Z}}$. Then n is ϵ-restrained if, for each $j \in \mathbf{Z}$, there exists $\omega = \omega_j$ such that $K(\omega, \epsilon)$ is defined and

$$\omega \in B_n[j - K(\omega, \epsilon), j + K(\omega, \epsilon)] .$$

We recall once more that $K(\omega, \epsilon)$ is defined *if ω is irrational or it is rational with denominator (in lowest terms) $> \epsilon^{-1}$*, as $K(\omega, \epsilon) = q_{\ell-1} + q_\ell$, where ℓ is the smallest integer such that $q_\ell > \epsilon^{-1}$ and $p_1/q_1, p_2/q_2, \ldots$ are the convergents of the continued fraction expansion of ω.

The proofs of Lemma 19.18 and Lemma 19.18' are elementary and follow from the definitions of $A_n[j, k]$, $B_n[j, k]$ and from what it means for n to be ϵ-restrained. Details can be found in [Ma11, Lemma 7.2-7.3]. It can also be shown that every ϵ-restrained biinfinite sequence of integers has a *unique* Farey interval associated to it in the above sense. However, what it really matters for our purposes is the possibility of constructing ϵ-restrained sequences which permit us to achieve *arbitrary sequences of rotation numbers* $(\ldots, \omega_j, \ldots)$, *as long as they are contained in a Farey interval* Ω. We skip the details of this construction which can be found in [Ma11, §7].

The solution of the number theoretical problem leads, in view of Lemma 19.17, to the possibility of constructing appropriate constraints $\mathcal{J}$ (i.e. constraints for which every $\mathcal{J}$-minimal configuration is $\mathcal{J}$-free) "following" an arbitrary sequence of rotation numbers $(\ldots, \omega_j, \ldots)$, *as long as these rotation numbers are contained in a Farey interval, whose heigth is determined by a lower bound for the Peierls's barriers corresponding to rotation symbols in that interval.* Furthermore, it can be proved, essentially as a consequence of the Aubry's Crossing Lemma, that, given $\epsilon > 0$, if the constraint $\mathcal{J}$ "follows" a minimal configuration x of rotation number $\omega \in \mathbf{R} \setminus \mathbf{Q}$ for a sufficiently long segment $(J_{i_0}, \ldots, J_{i_1})$, then the f-orbits corresponding to x and to any $\mathcal{J}$-minimal and $\mathcal{J}$-free configuration ξ approach within ϵ on a subsegment $i_0 + K \leq i \leq i_1 - K_1$, i.e.

$$|\xi_i - x_i| + |\eta_i - y_i| < \epsilon ,$$

where $(\ldots, (x_i, y_i), \ldots)$ and $(\ldots, (\xi_i, \eta_i), \ldots)$ are f-orbits corresponding to $x = (\ldots, x_i, \ldots)$ and $\xi = (\ldots, \xi_i, \ldots)$. An analogous result holds in case $\omega \in \mathbf{Q}$. We refer to [Ma11, §10] for details.

The above arguments lead to the following *partial* results in the direction of Theorems 19.12 and 19.13. Let $a \in \mathbf{R}$ and $P > 0$. Let $\Omega \subset \mathbf{R}$ be an open interval. Let k be the smallest integer $> 2\theta/P$ and suppose that *for any rational number $p/q \in \Omega$ (in lowest terms), we have $q > k$.* The positive number P has to be chosen to be a lower bound for the Peierls's barriers associated to rotation symbols in Ω, i.e. $P_\omega^h(a) > P$, for any $\omega \in \Omega$.

Proposition 19.19. *Consider $\omega_-, \omega_+ \in \Omega$. There exists an f-orbit $\mathcal{O}$ which is α-asymptotic to $\Sigma_{\omega_-}^\star$ and ω-asymptotic to $\Sigma_{\omega_+}^\star$.*

Proposition 19.20. *Consider for each $i \in \mathbf{Z}$, a real number $\omega_i \in \Omega$ and a positive number ϵ_i. There exists an f-orbit $\mathcal{O} = (..., (\theta_i, y_i), ...)$ and an increasing bi-infinite sequence $..., j(i), ...$ of integers such that*

$$dist\left((\theta_{j(i)}, y_{j(i)}), \Sigma_{\omega_i}^\star\right) < \epsilon_i \ .$$

In Proposition 19.19 and 19.20 it is summarized the construction of orbits which approach Aubry-Mather sets whose rotation numbers lie in an interval Ω. The limitation of these results consists in the fact that the interval Ω cannot contain any rational numbers p/q with $q \leq 2\theta/P$, where $P = \sup_{a \in \mathbf{R}} \inf_{\omega \in \Omega} \{P_\omega^h(a)\}$. If Ω contains a rational number with such a small denominator than the above results do not apply. However, it is possible to prove an analogous of Lemma 19.16 which allows to construct appropriate constraints $\mathcal{J}$ which "follow" sequences $(..., \omega_j, ...)$ of rotation numbers across any rational rotation number, independently of the size of its denominator, thereby completing the ingredients for the proof of Theorems 19.12 and 19.13.

Let p/q be a rational number, in lowest terms and with $q \geq 0$. It is possible to construct a segment of an orbit which first follows a minimal orbit of rotation number slightly less than (resp. slightly greater than) p/q, then follows a minimal orbit of rotation symbol p/q^- (resp. p/q^+), then follows a minimal orbit of rotation symbol p/q^+ (resp. p/q^-), and finally follows a minimal orbit of rotation symbol slightly greater than (resp. less than) p/q. Furthermore, it is possible to construct such an orbit so that it approaches arbitrarily closely the minimal periodic orbit of rotation symbol p/q. This results completes Lemma 19.16, which provides a segment of an orbit following a minimal orbit whose rotation symbol is irrational or is a rational number of large denominator. The proof of the result just described is long and quite technical, therefore we will omit it. We simply observe that it is based on the positivity of the Peierls's barriers corresponding to the rotation symbols $p/q^\pm$, in the same spirit of Lemma 19.16. Details can be found in [Ma11, Proposition 8.1].

§20. Action Minimizing Invariant Measures for Positive Definite Lagrangian Systems.

In this section we will discuss a generalization to more degrees of freedom of the variational approach to area-preserving mappings which represent the case of 2 degrees of freedom. We will not generalize the notion of minimal orbit, introduced in §9 and §14, but the related notion of *minimal measure*, which is a measure minimizing the action functional in sense to be specified (see [Ma10]). In generalizing to more degrees of freedom a major difficulty consists in finding the right setting. The setting proposed is inspired by a result due to Moser [Mo2], already mentioned in §15. According to this result, any exact area-preserving monotone twist map f of the annulus (or any finite composition of them) can be interpolated by a time dependent (non-autonomous) periodic Hamiltonian flow on T^*S^1, induced by a Hamiltonian $H : T^*S^1 \times \mathbf{R} \times \mathbf{R} \to \mathbf{R}$, satisfying the Legendre condition

$$H_{yy}(\theta, y, t) > 0 \ ,$$

i.e. f coincides with the time-1-map of the Hamiltonian flow associated to H. Since H satisfies the Legendre condition, f can also be interpolated by the time-1-map associated to the Lagrangian flow which can be obtained from the previous Hamiltonian flow by the usual Legendre transformation.

Thus, periodic positive definite Lagrangian systems provide a setting which generalize at once exact area-preserving twist mappings of the annulus and the geodesic flow on Riemannian manifolds diffeomorphic to the 2-torus. For these problems (2 degrees of freedom) a series of related results are known, namely the first author's results, exposed in §6 (and the closely related results by Aubry-Le Daeron [Au-LeD]) for area-preserving mappings and Hedlund results concerning "class A" geodesics of Riemannian metrics on the 2-torus [Hd]. The reader can consult on these subjects and their mutual connections the survey paper by V.Bangert [Ba]. In this section we will describe a generalization of these results, due to the first author [Ma12], to more degrees of freedom: an existence theorem for minimal measures, and a regularity theorem which asserts that the support of minimal measures can be expressed as a (partially defined) Lipschitz section of the tangent bundle. The first result generalizes Theorem 6.1, while the second extends to more degrees of freedom Theorem 14.1. The set of *ergodic* minimal invariant measures is also (partially) described for generic Lagrangians, following Mañé [Mñ]. All his results have been obtained by a slight modification of the setting proposed by the first author in [Ma12], which we will now describe.

Let M be a compact, connected C^∞ manifold, and TM be its tangent bundle. Let $L : TM \times \mathbf{R} \to \mathbf{R}$ be a C^2 function, called the "Lagrangian". The typical situation is when M is a torus. In particular, when $M = \mathbf{S}^1$, then $TM = \mathbf{S}^1 \times \mathbf{R}$ is an infinite cylinder, which is familiar in the theory of twist maps. We impose the following conditions on the Lagrangian L. We suppose that L is *periodic* with respect to the $\mathbf{R}$ coordinate, i.e. $L(\xi, t) = L(\xi, t+1)$, $\xi \in TM$ and $t \in \mathbf{R}$, where the period is, for convenience, normalized to be 1. We suppose that L has *positive definite* fiberwise Hessian everywhere, i.e. $L|TM_x$ has positive definite Hessian, for any $x \in M$. We suppose that L has fiberwise *superlinear growth*, i.e.

$$L(\xi, t)/\|\xi\| \to +\infty , \quad \text{as } \|\xi\| \to +\infty , \quad \text{for } \xi \in TM , t \in \mathbf{R} .$$

Here $\| \cdot \|$ denotes the norm associated to a Riemannian metric on M. Since M is compact, this condition does not depend on the choice of the Riemannian metric.

The last two conditions imply that the Legendre transformation $\mathcal{L}$ is defined: if $x \in M$, $v \in TM_x$, $t \in \mathbf{R}$, then

$$(20.1) \qquad \mathcal{L}(x, v, t) = (x, d_v(L|TM_x \times \{t\}), t) .$$

If L is C^r ($r \geq 2$), then $\mathcal{L}$ is a C^{r-1} diffeomorphism of $TM \times \mathbf{R}$ onto $TM^* \times \mathbf{R}$, where TM^* denotes the cotangent bundle of M.

The fourth condition regards the *completeness* of the Euler-Lagrange flow, associated to L. The Euler-Lagrange flow can be obtained by the the first variation of the action functional in the following way. We pose the variational problem for the functional

$$(20.2) \qquad A(\gamma) = \int_a^b L(d\gamma(t), t)\, dt$$

over C^1 curves $\gamma : [t_0, t_1] \to M$ with the fixed endpoints constraint. Here, $d\gamma$ denotes the differential of the map γ. The trajectories of the Euler-Lagrange flow correspond to the solution of the variational equation

$$(20.3) \qquad \delta A(\gamma) = 0 ,$$

associated to the variational problem for $A(\gamma)$ (with fixed endpoints). In other words, a C^1 curve in $TM \times \mathbf{S}^1$ is a trajectory of the Euler-Lagrange flow if and only if it is of the form $(d\gamma(t), t(\mathrm{mod}.\, 1))$, where γ is a curve on M which satisfies the variational equation (20.3). The first variation of the functional $A(\gamma)$ over the space of curves with fixed endpoints can be computed as:

$$(20.3') \qquad \delta A(\gamma)(\Gamma) = \frac{d}{ds} \left(\int L(\frac{\partial \Gamma}{\partial t}(s, t))\, dt \right) \Big|_{s=0} ,$$

for any C^1 mapping $\Gamma : [-\epsilon, \epsilon] \times [t_0, t_1] \to M$ such that $\Gamma(0, t) = \gamma(t)$, for all $t \in [t_0, t_1]$ and $\Gamma(s, t_0) = \gamma(t_0)$, $\Gamma(s, t_1) = \gamma(t_1)$, for all $s \in [-\epsilon, \epsilon]$. It is well known that (20.3), with respect to a system of C^∞ coordinates $(x_1, ..., x_n)$, takes the form:

$$(20.3'') \qquad \frac{d}{dt} L_{\dot{x}} = L_x .$$

Therefore, the Euler-Lagrange flow is associated to the vectorfield E_L described by

$$(20.4) \qquad \frac{dx}{dt} = \dot{x} , \quad \frac{d}{dt} L_{\dot{x}} = L_x .$$

The Euler-Lagrange vectorfield corresponds, through the Legendre transformation, to a Hamiltonian vectorfield on TM^*. It is not difficult to show that, if the Lagrangian L is C^r, the corresponding Hamiltonian function is also C^r, thus the

Hamiltonian vectorfield is C^{r-1}. Consequently, since the Legendre transformation is C^{r-1}, the Euler-Lagrange flow is C^{r-1}, although the Euler-Lagrange vectorfield (20.4) may be only C^{r-2}. Since $r \geq 2$, we obtain that even though the vectorfield E_L may be only C^0, it satisfies the conclusion of the fundamental existence and uniqueness theorem for ordinary differential equations. We now state the fourth condition. The Euler-Lagrange flow is *complete*, i.e. every maximal integral curve of the vectorfield E_L has all of $\mathbf{R}$ as its domain of definition.

In the classical calculus of variations the following basic result, concerning the above boundary value problem, holds:

Tonelli's Theorem. *Let $a < b \in \mathbf{R}$, and let x_a, $x_b \in M$. If $L : TM \times \mathbf{R} \to \mathbf{R}$, periodic with respect to the $\mathbf{R}$ coordinate, is fiberwise positive definite and has superlinear growth, then, among the absolutely continuous curves $\gamma : [a, b] \to M$ such that $\gamma(a) = x_a$ and $\gamma(b) = x_b$, there is one which minimizes the action*

$$A(\gamma) = \int_a^b L(d\gamma(t), t)\, dt .$$

As pointed out by Mañé in [Mñ], it is not necessary to assume compactness of M for the Tonelli's theorem to hold, if the superlinear growth condition is satisfied with respect to some complete Riemannian metric on M.

A curve which minimizes in the sense of Tonelli's theorem is called a *Tonelli minimizer*. Ball and Mizel [B-M] have constructed examples of Tonelli minimizers which are not C^1, under the hypotheses of Tonelli's Theorem. However, under the additional hypothesis of *completeness* of the Euler-Lagrange flow, a Tonelli minimizer γ must be C^1, and therefore it satifies the Euler-Lagrange equation. In case L is C^r, we have seen that a trajectory $t \to (d\gamma(t), t)$ of the Euler-Lagrange flow is C^{r-1}, thus γ is C^r. The role of the completeness hypothesis can be explained as follows. It is possible to prove that, under the hypotheses of Tonelli's Theorem, a minimizer γ not only exists and belongs to the space of absolutely continuous functions, but it is C^1 on an open and dense set of full measure in the interval in which it is defined and its velocity goes to the infinity on the exceptional set. Consequently, the completeness hypothesis implies that a Tonelli minimizer is C^1 (and hence C^r).

Let $\mathcal{M}_L$ be the space of E_L-invariant probability measures on $P = TM \times \mathbf{S}^1$. To every $\mu \in \mathcal{M}_L$, we may associate its *average action*

(20.5)
$$A(\mu) = \int_P L\, d\mu .$$

Since L is bounded below, the integral exists although it may be $+\infty$. In case $A(\mu) < +\infty$, we may associate to μ its *rotation vector* $\rho(\mu) \in H_1(M, \mathbf{R})$, which can be uniquely characterized as follows. Let $c \in H^1(M, \mathbf{R})$ be a cohomology class. By the de Rham Theorem, c can be represented by a closed 1-form λ. A differential 1-form is defined as a section of the cotangent bundle TM^*, but it can be considered also as a function on TM, linear on fibers, hence as a function on P.

Then, the integral on the right in the equation below is defined and it is is finite, since $A(\mu) < +\infty$ and L satisfies the superlinear growth condition (along fibers):

$$(20.6) \qquad (c, \rho(\mu)) = \int_P \lambda \, d\mu \ .$$

The bracket on the left denotes the canonical pairing between the cohomology group $H^1(M, \mathbf{R})$ and the homology group $H_1(M, \mathbf{R})$. It is elementary to show that, since μ is E_L-invariant, if λ is an exact form, then the integral on the rigth in (20.6) vanishes [Ma12, §2, Lemma]. Since this integral is linear with respect to $c \in H^1(M, \mathbf{R})$, (20.6) defines a homology class $\rho(\mu) \in H_1(M, \mathbf{R})$. The basic idea of rotation vector goes back to Schwartzman's *asymptotic cycles* [Sw].

It is not difficult to show the existence of invariant probability measures μ such that $A(\mu) < +\infty$, for which consequently the rotation vector $\rho(\mu)$ is defined. The argument is essentially based on the Kryloff-Bogoliuboff procedure to construct probability invariant measures for continuous flows on compact spaces. However, the space $P = TM \times \mathbf{S}^1$ is not compact. Therefore, we will consider the one point compactification $P^* = P \cup \{\infty\}$. The Euler-Lagrange flow easily extends to P^* to a flow which fixes ∞ and the Lagrangian L can be extended by $L(\infty) = \infty$ to a function $L : P^* \to \overline{\mathbf{R}}$.

Lemma 20.1. $A(\mu) = \int L \, d\mu$ *is a lower-semicontinuous functional on the space of Borel probability measure on P^* with the vague (weak) topology. Furthermore, there exists $\mu \in \mathcal{M}_L$ such that $A(\mu) < +\infty$.*

Proof. Let $A_K(\mu) = \int \min(L, K) \, d\mu$, for $K \in \mathbf{R}$. Then A_K is continuous, since $\min(L, K)$ is a bounded function, and $A_K(\mu) \nearrow A(\mu)$, as $K \nearrow +\infty$. This implies the lower semicontinuity of $A(\mu)$.

We now apply the Kryloff-Bogoliuboff argument. Let α_n be an absolute minimizer (i.e. with free boundaries) defined on a time interval of length n. Let $\gamma_n(t) = (d\alpha_n(t), t)$. By the previous remarks, γ_n is a trajectory of the Euler-Lagrange flow. Let μ_n be the probability measure evenly distributed along γ_n and let μ be an accumulation point of the set $\{\mu_n\}_{n \in \mathbf{N}}$, with respect to the vague topology on the space of the Borel probability measures on P^*. This exists because P^* is a compact space. An elementary argument, which we will omit, shows that μ is an invariant measure for the extended Euler-Lagrange flow. On the other hand, it clearly exists for each $n \in \mathbf{N}$ some curve β_n, defined on an interval of length n, such that $A(\beta_n) \leq C \, n$. Hence,

$$(20.7) \qquad A(\mu_n) = n^{-1} A(\alpha_n) \leq A(\beta_n) \leq C \ .$$

Therefore $A(\mu) \leq C$, by the lower semicontinuity of the action functional on the space of probability measures. Finally, since $L(\infty) = \infty$ and $A(\mu) < \infty$, the measure μ just constructed has no atomic part supported at the fixed point ∞. Hence its restriction to P is a probability measure on P and $\mu \in \mathcal{M}_L$. $\qquad \square$

The lemma we just proved has the following immediate consequence: *there exists $\mu \in \mathcal{M}_L$ which minimizes A over $\mathcal{M}_L$.*

A refinement of the previous Kryloff-Bogoliuboff argument gives the following:

Lemma 20.2. *Let $h \in H_1(M, \mathbf{R})$. Then there exists $\mu \in \mathcal{M}_L$ satisfying $A(\mu) < +\infty$ and $\rho(\mu) = h$.*

Proof. We will apply Tonelli's Theorem to the covering space $\tilde{M}$ of M, determined by $\pi_1(\tilde{M}) = \ker(\mathcal{H} : \pi_1(M) \to H_1(M, \mathbf{R}))$, where $\mathcal{H}$ denotes the Hurewicz homomorphism. In the model case $M = \mathbf{T}^n$, then $\tilde{M} = \mathbf{R}^n$. The group of deck transformations of this covering space can be identified to

$$(20.8) \quad \mathcal{D} = \mathrm{Im}\,(\mathcal{H} : \pi_1(M) \to H_1(M, \mathbf{R})) = \mathrm{Im}\,(\mathcal{H} : H_1(M, \mathbf{Z}) \to H_1(M, \mathbf{R}))\,,$$

which clearly is a lattice in the finite dimensional vector space $H_1(M, \mathbf{R})$. For example, if $M = \mathbf{T}^n$, then $\mathcal{D} = \mathbf{Z}^n$.

Therefore, for any $h \in H_1(M, \mathbf{R})$, there exists a sequence of deck transformations $T_1, ..., T_n, ...$ such that $n^{-1} T_n \to h \in H_1(M, \mathbf{R})$, as $n \to +\infty$. Fix $\tilde{x}_0 \in \tilde{M}$ and let $\tilde{x}_n = T_n \tilde{x}_0$. Let $\tilde{\alpha}_n : [0, n] \to \tilde{M}$ minimize $\int_0^n L(d\alpha_n(t), t)\, dt$, subject to the boundary conditions $\tilde{\alpha}_n(0) = \tilde{x}_0$ and $\tilde{\alpha}_n(n) = \tilde{x}_n$, where α_n is the projection of $\tilde{\alpha}_n$ on M. The existence of $\tilde{\alpha}_n$ follows from an adapted version of Tonelli's Theorem (cf. [Ma12, §2]). To obtain a E_L-invariant measure, we apply again the Kryloff-Bogoliuboff argument. By the completeness hypothesis of the Euler-Lagrange flow $\tilde{\alpha}_n$ is C^1. We let $\gamma_n = (d\alpha_n(t), t) \in TM \times S^1$. We let μ_n be the probability measure evenly distributed along γ_n and let μ be an accumulation point of the set $\{\mu_n\}_{n \in \mathbf{N}}$, with respect to the weak topology on the space of Borel probabiliy measures on P^*. As before, it is elementary to show that μ is an invariant measure for the Euler-Lagrange flow. This is the core of the Kryloff-Bogoliuboff argument. On the other hand it is easy to see that there exists $C > 0$ and, for each $n \in \mathbf{N}$, a curve $\tilde{\beta}_n : [0, n] \to \tilde{M}$ such that $\tilde{\beta}_n(0) = \tilde{x}_0$, $\tilde{\beta}_n(n) = \tilde{x}_n$ and $\int_0^n L(d\beta_n(t), t)\, dt \leq Cn$, where β_n is the projection of $\tilde{\beta}_n$ on $\tilde{M}$. Consequently,

$$(20.9) \quad A(\mu_n) = n^{-1} A(\alpha_n) \leq A(\beta_n) \leq C\,,$$

which clearly implies $A(\mu) < +\infty$. In particular the point at infinity has zero measure with respect to μ, thus μ can be viewed as a probability measure on $TM \times S^1$, invariant with respect to the Euler-Lagrange flow. Finally, by construction

$$(20.10) \quad \rho(\mu) = \lim_{n \to +\infty} n^{-1} T_n = h \in H_1(M, \mathbf{R})\,.$$

This complete the outline of the proof of Lemma 20.2. $\qquad \square$

It follows immediately from the conditions imposed on L that L is bounded from below, i.e. there exists $B \in \mathbf{R}$ such that $L \geq B$. Therefore the set $U_L = \{(\rho(\mu), z) \in H_1(M, \mathbf{R}) \times \mathbf{R} \,|\, A(\mu) \leq z, \ \mu \in \mathcal{M}_L\}$ is contained in $H_1(M, \mathbf{R}) \times [B, +\infty)$. Furthermore, since $\mu \to \rho(\mu)$ is continuous on $\{\mu \in \mathcal{M}_L \,|\, A(\mu) \leq C\}$, as it is not difficult to prove, and A is lower semicontinuous (Lemma 20.1), U_L is a closed set. Clearly, U_L is also convex and, by Lemma 20.2, its projection on $H_1(M, \mathbf{R})$ is surjective. Consequently, U_L is the *epigraph of a convex function* $\beta = \beta_L : H_1(M, \mathbf{R}) \to \mathbf{R}$, i.e. $U_L = \{(h, z) \in H_1(M, \mathbf{R}) \times \mathbf{R} \,|\, \beta(h) \leq z\}$. For any $h \in H_1(M, \mathbf{R})$, we will call $\beta(h)$ the *minimal average action* of the rotation vector h.

The value $\beta(h)$ clearly represents the minimum of the average action A over the probability measures $\mu \in \mathcal{M}_L$ satisfying $\rho(\mu) = h$. This is a minimization problem for the functional A, subject to the contraint $\rho(\mu) = h$, and can be treated by the method of Lagrange multipliers. This remark motivates the introduction of the minimization problem for the following family of functionals. Let $c \in H^1(M, \mathbf{R})$, we set

$$(20.11) \qquad A_c(\mu) = A(\mu) - (c, \rho(\mu)) = \int_P (L - \lambda) \, d\mu \ ,$$

where λ is a closed 1-form in the de Rham cohomology class corresponding to c. This is defined for any $\mu \in \mathcal{M}_L$ such that $A(\mu) < +\infty$ and it can be extended to the case $A(\mu) = +\infty$, by setting $A_c(\mu) = +\infty$. The cohomology class c plays the role of the Lagrange multiplier. $L - \lambda$ satisfies the same conditions imposed on L, in particular the Euler-Lagrange flow associated to $L - \lambda$ is the same as that of L. Consequently, Lemma 20.1 applies and A_c takes a minimum value, which will be denoted by $-\alpha(c)$. It is not difficult to realize that $\alpha : H^1(M, \mathbf{R}) \to \mathbf{R}$ is a convex function and its epigraph $\{(c, z) \in H^1(M, \mathbf{R}) \times \mathbf{R} \,|\, z \geq \alpha(c)\}$ is a convex subset of $H^1(M, \mathbf{R}) \times \mathbf{R}$.

Let $\alpha^* : H_1(M, \mathbf{R}) \to \mathbf{R}$ denote the *conjugate* function of α in the sense of convex analysis [Rc]:

$$(20.12) \qquad -\alpha^*(h) = \min\{\alpha(c) - (c, h)\} \ ,$$

where c varies over $H^1(M, \mathbf{R})$. A priori, α^* takes values in $\overline{\mathbf{R}} = \mathbf{R} \cup \{+\infty\}$, but in fact it takes values in $\mathbf{R}$. The reason is the following: if $\mu \in \mathcal{M}_L$ and $A(\mu) < +\infty$, then $\alpha^*(\rho(\mu)) \leq A(\mu)$, as it follows immediately from the definitions. On the other hand, by Lemma 20.2, for any $h \in H_1(M, \mathbf{R})$, there exists $\mu \in \mathcal{M}_L$ such that $A(\mu) < +\infty$ and $\rho(\mu) = h$. Furthermore, by its definition, $\alpha = \beta^*$. The basic convex analysis then implies:

Theorem 20.3. [Ma12, Theorem 1] *The functions $\alpha : H^1(M, \mathbf{R}) \to \mathbf{R}$ and $\beta : H_1(M, \mathbf{R}) \to \mathbf{R}$ are conjugate convex functions and have superlinear growth. For $h \in H_1(M, \mathbf{R})$, we have*

$$\beta(h) = \min\{A(\mu) \,|\, \mu \in \mathcal{M}_L \quad \text{and} \quad \rho(\mu) = h\} \ .$$

For $c \in H^1(M, \mathbf{R})$, we have

$$-\alpha(c) = \min\{A_c(\mu) \,|\, \mu \in \mathcal{M}_L\} \ .$$

The outcome of the above discussion can be summarized in the following terms. Let $\mathcal{M}^c$, $c \in H^1(M, \mathbf{R})$, be the set of all probability measures $\mu \in \mathcal{M}_L$ which minimize A_c and $\mathcal{M}_h$ the set ot all $\mu \in \mathcal{M}_L$ which minimize A and satisfy $\rho(\mu) = h$. Then

$$(20.13) \qquad \bigcup_{c \in H^1(M,\mathbf{R})} \mathcal{M}^c = \bigcup_{h \in H_1(M,\mathbf{R})} \mathcal{M}_h ,$$

in fact, if $\mu \in \mathcal{M}_L$, then $A(\mu) = \beta(\rho(\mu))$ if and only if there exists $c \in H^1(M, \mathbf{R})$ such that μ minimizes A_c. Furthermore, c is the subderivative of the convex function β at $\rho(\mu)$, i.e. the slope of a supporting hyperplane of the epigraph of β at $\rho(\mu)$. A probability measure μ satisfying these conditions is called a *minimal measure*.

The invariant probability measures which are relevant for the dynamics are the *ergodic* ones, i.e. those satisfying the condition that every invariant set has measure 0 or 1. It is a standard elementary result of topological dynamics that the extremal points of $\mathcal{M}_L$ are the ergodic measures invariant with respect to the Euler-Lagrange flow. Since the convex function β has superlinear growth, according to Theorem 20.3, its epigraph has infinitely many extremal points by a standard result in convex analysis [Rc]. Let $(h, \beta(h))$ denote an extremal point of the epigraph of β. The extremal points of the set of $\mu \in \mathcal{M}_L$ for which $\rho(\mu) = h$ and $A(\mu) = \beta(h)$ are ergodic measures, since they also are extremal points of $\mathcal{M}_L$. Since this set is compact and convex such extremal points do exist. Thus, we have proved the existence of *at least one* invariant ergodic minimal measure μ with rotation vector h, such that $(h, \beta(h))$ is an extremal point of the epigraph of β. This result opens the problem of describing the set of rotation vectors h for which $\mathcal{M}_h$ contains an ergodic measure.

The simplest case to understand is when $M = \mathbf{S}^1$. It follows from a theorem due to Moser [Mo2] that this case is related to twist maps or, more generally, to finite compositions of twist map. In fact, as we already mentioned at the beginning of this section, according to this result, any finite composition of exact area-preserving monotone twist maps f of the annulus can be interpolated by a time dependent (non-autonomous) periodic Lagrangian flow on $T\mathbf{S}^1$, induced by a Lagrangian $L : T\mathbf{S}^1 \times \mathbf{R} \rightarrow \mathbf{R}$ which satisfies the conditions imposed here, i.e. the positive definiteness, the superlinear growth (along fibers) and the completeness of the Euler-Lagrange flow. In this new setting the existence theorem of Aubry-Mather sets (Theorem 6.1) for twist maps (or finite compositions of them) is a consequence of the following:

Proposition 20.4. *In the case* $M = \mathbf{S}^1$, *the function* $\beta : H_1(M, \mathbf{R}) \rightarrow \mathbf{R}$ *is strictly convex, i.e. every point on the graph of* β *is an extremal point of the epigraph of* β. *Therefore,* $\mathcal{M}_h$ *contains an ergodic measure, for any* $h \in H_1(M, \mathbf{R})$.

This result depends on the Lipschitz property of the support of minimizing measures, which will be described later.

We will outline below some results obtained by Mañé [Mñ] in the direction of giving a satisfactory description of the set of ergodic minimal measures in the general case, at least for generic Lagrangians. These results leave many natural questions unsolved and the situation is far from being as simple and well understood as in the case $M = \mathbf{S}^1$. Mañé approach is inspired to the first author's approach described above, but it differs slightly under the following respect. The minimizing measures are obtained through a variational principle not requiring the invariance a priori, but the invariance property is proved as a consequence of the minimization property over an appropriate space.

Let $\mathcal{M}$ be the set of probability measures on the Borel σ-algebra of $TM \times \mathbf{S}^1$ such that:

(20.14)
$$\int_P \|\xi\| \, d\mu \; < \; +\infty \; .$$

It is not difficult to show (cf. [Mñ]) that there exists a unique metrizable topology on $\mathcal{M}$ such that

(20.15)
$$\mu_n \to \mu \iff \int \psi \, d\mu_n \to \int \psi \, d\mu \; ,$$

for every continuous function $\psi : TM \times \mathbf{S}^1 \to \mathbf{R}$ growing (fiberwise) *at most linearly*, i.e.

(20.16)
$$\sup_{(x,\xi,t)} \frac{\psi(x,\xi,t)}{1 + \|\xi\|} \; < \; +\infty \; .$$

The space of probability measures which will be considered is a closed subset of $\mathcal{M}$ (endowed with the induced topology) defined as follows. Given a periodic absolutely continuous curve $\gamma : \mathbf{R} \to M$, with period $N \in \mathbf{Z}$, define the probability μ_γ on the Borel σ-algebra of $TM \times \mathbf{S}^1$ by posing

(20.17)
$$\int \psi \, d\mu_\gamma \; = \; \frac{1}{N} \int_0^N \psi(d\gamma(t), t) \, dt \; ,$$

for every continuous function $\psi : TM \times \mathbf{S}^1 \to \mathbf{R}$ with compact support. It is an immediate consequence of the absolute continuity of γ that $\mu_\gamma \in \mathcal{M}$, since (20.14) clearly holds if γ is absolutely continuous. Furthermore, observe that $\mu_{\gamma_1} = \mu_{\gamma_2}$ if and only if $\gamma_1 = \gamma_2$. A probability measure μ_γ has a naturally associated homology class $\tilde{\rho}(\mu_\gamma)$ defined as

$$\tilde{\rho}(\mu_\gamma) = \frac{1}{N} \, [\gamma] \in H_1(M, \mathbf{R}) \; ,$$

where $[\gamma]$ denotes the homology class of the curve γ. Let $\mathcal{C} \subset \mathcal{M}$ be the set of probability measures of the form μ_γ, and let $\overline{\mathcal{C}}$ be its closure. The space $\overline{\mathcal{C}}$ satisfies the following three properties:

(I) For every Lagrangian L on M, satisfying the conditions of positive definiteness, superlinear growth (along fibers) and completeness of the associated Euler-Lagrange flow, all the probability measures μ which are invariant with respect to the Euler-Lagrange flow such that

$$\int_P L \, d\mu \; < \; +\infty$$

are contained in $\overline{\mathcal{C}}$.

(II) Probabilities μ in $\overline{C}$ have a naturally associated homology $\tilde{\rho}(\mu) \in H_1(M, \mathbf{R})$. The map $\tilde{\rho} : \overline{C} \to H_1(M, \mathbf{R})$ is the continuous extension of the map $\tilde{\rho} : C \to H_1(M, \mathbf{R})$, defined above. Furthermore, the map ρ, extended to $\overline{C}$, is surjective.

(III) For every Lagrangian L on M, satisfying the above conditions, the set $\{\mu \in \mathcal{M} \mid \int L\, d\mu < +\infty\}$ is compact.

These properties are proved in [Mñ, §1]. It follows that the variational problem of finding, given a rotation vector $h \in H_1(M, \mathbf{R})$, a probability measure $\mu \in \overline{C}$ satisfying

$$\int_P L\, d\mu = \min\left\{ \int_P L\, d\nu \mid \nu \in \overline{C}, \ \tilde{\rho}(\nu) = h \right\}$$

has at least one solution. This is a consequence of the compactness property (III) and the continuity of the map $\tilde{\rho}$. We will denote the set of such μ by $\tilde{\mathcal{M}}_h$, in analogy with the previous notation. As before, if λ is a closed 1-form, the above assertions hold true for the Lagrangian $L - \lambda$. Thus, it is possible to consider the set $\tilde{\mathcal{M}}^c$ of measures $\mu \in \overline{C}$ which minimizes over $\overline{C}$ the average action functional associated to the Lagrangian $L - \lambda$. Here, as before, c denotes the (de Rham) cohomology class of the closed 1-form λ. Standard convex analysis [Rc] gives:

IV)
$$\bigcup_{c \in H^1(M, \mathbf{R})} \tilde{\mathcal{M}}^c = \bigcup_{h \in H_1(M, \mathbf{R})} \tilde{\mathcal{M}}_h .$$

A probability measure $\mu \in \overline{C}$, belonging to the set in (IV), will be called a $\overline{C}$-*minimal measure* for the Lagrangian L. The following result unifies the first author's approach and Mañé's approach, which we just described.

Theorem 20.5. *[Mñ, Theorem A] Any $\overline{C}$-minimal measure μ for a Lagrangian L, satisfying the stated conditions of positive definiteness, superlinear growth and completeness, is invariant under the Euler-Lagrange flow associated to L. Furthermore, if $\mu \in \tilde{\mathcal{M}}^c$ and λ is a 1-form in the cohomology class $c \in H^1(M, \mathbf{R})$, the function $L - \lambda$ is homologous to a constant on the support of μ, in the sense that there exists a Lipschitz function $V : \operatorname{supp}(\mu) \to \mathbf{R}$ and a costant $C > 0$ such that on $\operatorname{supp}(\mu)$*

$$L - \lambda = C + E_L V ,$$

where $E_L V$ the directional derivative of the function V with repect to the vectorfield E_L, which generates the Euler-Lagrange flow.

In view of this result, $\tilde{\mathcal{M}}_h = \mathcal{M}_h$, $h \in H_1(M, \mathbf{R})$, $\tilde{\mathcal{M}}^c = \mathcal{M}^c$, $c \in H^1(M, \mathbf{R})$, and the identities (20.13) and (IV) coincide. Thus, a $\overline{C}$-*minimal measure* in Mañé's sense is a *minimal measure* in the first author's sense. Furthermore, the function $\tilde{\rho} : \overline{C} \to H_1(M, \mathbf{R})$, which associate to each probability measure $\mu \in \overline{C}$ its rotation vector, coincides on the space $\mathcal{M}_L$ of probability measures invariant with respect to the Euler-Lagrange flow, with the map ρ introduced by the first author, characterized in (20.6). In fact, if μ is an ergodic invariant measures, it is a consequence of Birkhoff's ergodic theorem and of the continuity of $\tilde{\rho}$ that $\tilde{\rho}(\mu) = \rho(\mu)$ and it is uniquely determined as follows: let $\gamma : \mathbf{R} \to M$ be a μ-generic trajectory of the Euler-Lagrange flow (in the sense of the ergodic theorem). For $T > 0$, let z_T be the closed curve defined by $\gamma\|[-T, T]$ and a "short" curve (of length bounded by diam (M)) joining $\gamma(-T)$ with $\gamma(T)$. Then

(20.18)
$$\tilde{\rho}(\mu) = \rho(\mu) = \lim_{T \to +\infty} \frac{1}{2T} \left[z_T \right] .$$

We will state below the basic properties of minimal measures which have been established in [Ma12, §3]. Let $A_L \subset TM \times S^1$ be the union of all *minimizing trajectories* of the Euler-Lagrange flow, i.e. the set of all absolutely continuous (in fact smooth) curves $\gamma : [a, b] \to M$ which minimize the action

$$A(\gamma) = \int_a^b L(d\gamma(t), t) \, dt$$

over all absolutely continuous curves having the *same endpoints* $\gamma(a)$ and $\gamma(b)$ and belonging to the *same homology class* as γ in $H_1(M, \mathbf{R})$. A trajectory contained in A_L will also be called a *minimizer* for the Lagrangian L. A_L is clearly a closed invariant set. As before, $\mathcal{M}^c$, $c \in H^1(M, \mathbf{R})$, will denote the set of probability invariant measures minimizing A_c (20.11). As we remarked, $\mathcal{M}^c = \tilde{\mathcal{M}}^c$, where the latter is the set of measures in $\overline{\mathcal{C}}$ minimizing A_c constructed in the Mañé's approach. Furthermore, $\mathcal{M}^c$ is a compact, convex set and its extremal points are ergodic measures. We will denote by $\operatorname{supp} \mathcal{M}^c$ the *support* of $\mathcal{M}^c$, i.e. the set of $(x, \xi, t) \in TM \times S^1$ such that every neighborhood of (x, ξ, t) has positive measure with respect to some measure $\mu \in \mathcal{M}^c$.

Proposition 20.6. *[Ma12, Prop. 3] For any $c \in H^1(M, \mathbf{R})$, every trajectory of the Euler-Lagrange flow in $\operatorname{supp} \mathcal{M}^c$ is a minimizer, i.e. $\operatorname{supp} \mathcal{M}^c$ is contained in A_L.*

Proposition 20.7. *[Ma12, Prop. 2] If $\mu \in \mathcal{M}_L$ is ergodic and $\operatorname{supp}(\mu) \subset A_L$, then μ is a minimizing measure, i.e. there exists $c \in H^1(M, \mathbf{R})$ such that $\mu \in \mathcal{M}^c$.*

In view of the above two Propositions, it is clear that minimizing measures completely describe the ergodic theory of the invariant set A_L of all minimizing trajectories. It may seem that the dynamics on such an invariant set is restricted to a special behaviour. However, as pointed out in [Mñ], the dynamics on A_L can be as complicated as that of any vectorfield X on M. In fact, if we consider the autonomous Lagrangian

(20.19)
$$L(x, \xi) = \|\xi - X(x)\|^2 ,$$

then every every solution of the differential equation $\dot{x} = X(x)$ on M is a *minimizing* trajectory of the Euler-Lagrange flow associated to L, as the reader can easily verify.

Proposition 20.8. *[Ma12, Props. 2-3-4] $\operatorname{supp} \mathcal{M}^c$ is a compact subset of $TM \times S^1$. Furthermore, for any measure $\mu \in \mathcal{M}_L$,*

$$\operatorname{supp}(\mu) \subset \operatorname{supp} \mathcal{M}^c \implies \mu \in \mathcal{M}^c .$$

The arguments employed in the proof of the above Propositions 20.6 and 20.7 depends essentially on cutting and pasting minimizing trajectories and using the continuity of $\rho(\mu)$ and the lower semicontinuity of $A(\mu)$ combined with the Kryloff-Bogoliuboff procedure of constructing invariant measures. In the proof of Proposition 20.8, the main point is that, if $\|\xi\|$ is unbounded as (x, ξ, t) varies over $TM \times S^1$, then it is possible to construct an incomplete (i.e. not C^1) trajectory of the Euler-Lagrange flow, in contradiction with our hypotheses.

The next property we will state is the Lipschitz property of the support of $\mathcal{M}^c$, which is the main result of [Ma12]:

Theorem 20.9. *[Ma12, Th. 2] If $\pi : TM \times S^1 \to M \times S^1$ is the canonical projection, then, for any $c \in H^1(M, \mathbf{R})$, the restriction $\pi \,|\, supp\,\mathcal{M}^c$ is injective and its inverse $(\pi \,|\, supp\,\mathcal{M}^c)^{-1} : \pi(supp\,\mathcal{M}^c) \to supp\,\mathcal{M}^c$ is a Lipschitz map, i.e. there exists a constant $C > 0$ such that, for any x, $y \in \pi(supp\,\mathcal{M}^c)$, we have*

$$dist\,(\pi^{-1}(x), \pi^{-1}(y)) \leq C\,dist\,(x, y) \ .$$

The intuitive idea of the proof of Theorem 20.9 is the following. There is a well known "curve shortening" lemma in basic Riemannian geometry which goes as follows. Let α and β be curves on a Riemannian manifold joining points P, P' and Q, Q' resp. Suppose that α and β cross. Then there exist curves a, joining P and Q', and b, joining Q and P', such that

$$(20.20) \qquad \text{length}\,(a) + \text{length}\,(b) \ < \ \text{length}\,(\alpha) + \text{length}\,(\beta) \ .$$

A similar "shortening lemma" holds for the action functional associated to a Lagrangian L. In fact, the following holds [Ma12, §4, Lemma]: if $K > 0$, then there exist constants ϵ, δ, η, $C > 0$ such that, if α, $\beta : [t_0 - \epsilon, t_0 + \epsilon] \to M$ are trajectories of the Euler-Lagrange flow, with $\alpha(t_0 - \epsilon) = P$, $\alpha(t_0 + \epsilon) = P'$ and $\beta(t_0 - \epsilon) = Q$, $\beta(t_0 + \epsilon) = Q'$, $\|d\alpha(t_0)\|$, $\|d\beta(t_0)\| \leq K$, $dist\,(\alpha(t_0), \beta(t_0)) \leq \delta$, and $dist\,(d\alpha(t_0), d\beta(t_0)) \geq C\,dist\,(\alpha(t_0), \beta(t_0))$, then there exists C^1 curves a, $b : [t_0 - \epsilon, t_0 + \epsilon] \to M$ such that $a(t_0 - \epsilon) = P$, $a(t_0 + \epsilon) = Q'$ and $b(t_0 - \epsilon) = Q$, $b(t_0 + \epsilon) = P'$, and

$$(20.20') \qquad A(\alpha) + A(\beta) - A(a) - A(b) \geq \eta\,dist\,(d\alpha(t_0), d\beta(t_0))^2 \ .$$

Thus, if π were not injective on $supp\,\mathcal{M}^c$, or its inverse were not Lipschitz, it would be possible to construct a probability measure $\mu \in \mathcal{M}_L$ for which $A_c(\mu) < A_c(\mathcal{M}^c)$, contradicting the definition of $\mathcal{M}^c$. This result would be achieved by "cutting and pasting" trajectories using the "curve shortening" lemma and the Tonelli's Theorem. Then the Kryloff-Bogoliuboff argument would provide the required measure, because of the continuity of $\rho(\mu)$ and the lower semicontinuity of $A(\mu)$. The details of the arguments sketched above can be found in [Ma12, §§3-4].

The first applications of Theorem 20.9 are to the description of the case $M = S^1$, thereby completing the picture given by Proposition 20.4 and re-obtaining the basic results found in §14.

Proof of Proposition 20.4

Suppose $\beta : H_1(\mathbf{S}^1, \mathbf{R}) \equiv \mathbf{R} \to \mathbf{R}$ is not strictly convex. Then the graph of β intersects a line l in $\mathbf{R}^2$ in a segment I not reduced to a point. Let $(h_0, \beta(h_0))$ and $(h_1, \beta(h_1))$ be the endpoints of I. These points are extremal points of the epigraph of β, hence there exist action minimizing ergodic measures μ_0 and μ_1 whose rotation number is h_0 resp. h_1. Each of them is contained in $\mathcal{M}^c$, where $c \in H^1(\mathbf{S}^1, \mathbf{R}) \equiv \mathbf{R}$ is the slope of the line l. By Theorem 20.9, the projection π of $\operatorname{supp} \mathcal{M}^c$ on $\mathbf{S}^1 \times \mathbf{S}^1$ is injective. But this contradicts the fact that two Birkhoff generic orbits γ_0, γ_1 in $\pi(\operatorname{supp}(\mu_0))$ resp. $\pi(\operatorname{supp}(\mu_1))$ must cross, since they have different rotation numbers. On the other hand, they are the projections of distinct (and therefore disjoint) trajectories of the Euler-Lagrange flow on $TS^1 \times \mathbf{S}^1$. $\qquad\square$

Let $h \in H_1(\mathbf{S}^1, \mathbf{R})$, let $l \subset H^1(\mathbf{S}^1, \mathbf{R}) \times \mathbf{R}$ be a supporting hyperplane of the epigraph of the function β, which passes through $(h, \beta(h))$ and let $c \in H^1(\mathbf{S}^1, \mathbf{R})$ be the slope of l. Let $M_h = TS^1 \times \{0\} \cap \operatorname{supp} \mathcal{M}^c$. By the strict convexity of β (Proposition 20.4) M_h is well defined (i.e. it is independent of the choice of l), since in this case $\mathcal{M}^c = \mathcal{M}_h$. The set M_h is a closed invariant set for the time-1 Poincaré map $f_L : TS^1 \to TS^1$ associated to the Euler-Lagrange flow. These maps include "twist mappings" defined in §2, as a particular case.

Corollary 20.10. *The projection π_1 of M_h ($\subset TS^1$) on $\mathbf{S}^1$ is injective and the inverse $\pi_1^{-1} : \pi_1(M_h) \to M_h \subset TS^1$ is Lipschitz.*

Corollary 20.10 includes Thorem 14.1, now re-obtained as an immediate consequence of Theorem 20.9.

Let $\pi : \mathbf{R}^2 \to \mathbf{S}^1 \times \mathbf{R} = TS^1$ be the standard projection and let $\tilde{M}_h = \pi^{-1}(M_h) \subset \mathbf{R}^2$. Let $\pi_1 : \mathbf{R}^2 \to \mathbf{R}$ be the projection on the first factor and let $\tilde{f}_L$ denote a lift of f_L to the universal cover $\mathbf{R}^2$. Since, by Corollary 20.10, the projection $\pi_1 : \tilde{M}_h \to \mathbf{R}$ is injective, $\tilde{M}_h$ inherits an order structure from that on $\mathbf{R}$.

Corollary 20.11. *$\tilde{f}_L : \tilde{M}_h \to \tilde{M}_h$ is order preserving. Consequently, if h is irrational, M_h supports a unique invariant measure μ_h, which is the unique minimal measure of rotation number h.*

Proof. The order preserving property follows immediately from the injectivity of the projection of $\operatorname{supp} \mathcal{M}^c$ on $\mathbf{S}^1 \times \mathbf{S}^1$, which is the content of Theorem 20.9 in the case $M = \mathbf{S}^1$. The unique ergodicity of the closed invariant set M_h is a standard consequence of the order preserving property. The proof is the same as in the case of an order preserving homemorphism of the circle of irrational rotation number. Finally, since all minimal measures of rotation number h are supported in M_h, it follows that there is a unique measure μ_h having such properties. $\qquad\square$

A different application of Theorem 20.9, to small perturbations of symplectic diffeomorphisms having an invariant torus, can be found in [Ma12, §5]. There, it is exploited the remark that we can use the Lipschitz property asserted by Theorem 20.9 (and the *a priori* bound on the Lipschitz constant which can be obtained through it) to *localize* the invariant set supp $\mathcal{M}^c$. The result can be stated as follows. Let f be a symplectic diffeomorphism of the symplectic manifold (N, ω), i.e. $f^*\omega = \omega$, where N is a $2n$-dimensional manifold and ω is a closed non-degenerate 2-form on N. A K.A.M invariant torus of f is an n-dimensional submanifold $\mathcal{T}$ of N such that $f(\mathcal{T}) = \mathcal{T}$ and $f|\mathcal{T}$ is smoothly conjugate to an irrational translation on the n-torus T^n by a vector $\rho = (\rho_1, ..., \rho_n) \in \mathbf{R}^n$ which satisfies a Diophantine condition, i.e. there exist contants $C, \beta > 0$ such that

$$(20.21) \qquad |k_0 + k_1\rho_1 + ... + k_n\rho_n| \geq C\left(|k_1| + ... + |k_n|\right)^{-\beta},$$

for all $k = (k_1, ..., k_n) \in \mathbf{Z}^n \setminus \{0\}$.

Proposition 20.12. *There exists $c_0 \in H^1(T^n, \mathbf{R}) = \mathbf{R}^n$ (which is the derivative at ρ of the function β associated to the Lagrangian system obtained by interpolating f in a neighborhood of the invariant K.A.M. torus $\mathcal{T}$), such that the following holds. If c is close enough to c_0 in $H^1(T^n, \mathbf{R}) = \mathbf{R}^n$ and g is sufficiently close to f in the C^1 topology on Hamiltonian perturbations of f, then there exists a g-invariant set $\mathcal{T}_c$ associated to c (which is supp $\mathcal{M}^c$ for the associated Lagrangian system). Furthermore, $\mathcal{T}_c$ is a Lipschitz graph over $\mathcal{T}$ and it converges in the Hausdorff topology to $\mathcal{T}$ as c tends to c_0 in $\mathbf{R}^n$ and g tends to f in the C^1 topology. Finally, if g has a K.A.M. torus sufficiently C^1-close to $\mathcal{T}$, then that torus is one of the sets $\mathcal{T}_c$.*

We will now turn to the existence results of ergodic minimal measures for generic Lagrangian systems obtained by Mañé [Mñ] following the approach sketched above. In Mañé's results a central role is played by the following class of minimal measures:

A measure $\mu \in \bar{\mathcal{C}}$ will be said a *uniquely minimal* measure of a Lagrangian L if $\mathcal{M}^c(L) = \{\mu\}$, for some $c \in H^1(M, \mathbf{R})$.

As a consequence of Proposition 20.8 uniquely minimal measures are uniquely ergodic (i.e. if μ is a uniquely minimal measure then $\nu = \mu$ for every probability measure μ such that supp $(\nu) \subset$ supp (μ)). In particular, μ is ergodic.

Theorem 20.13. *[Mñ, Th.B] Let L be a Lagrangian satisfying the hypotheses of positive definiteness, superlinear growth (along fibers) and completeness of the Euler-Lagrange flow. Then:*

a) For every $c \in H^1(M, \mathbf{R})$ there exists a residual subset $\mathcal{A}(c) \subset C^\infty(M \times \mathbf{S}^1)$ such that $\psi \in \mathcal{A}(c)$ implies card $\mathcal{M}^c(L + \psi) = 1$.

b) There exist residual subsets $\mathcal{A} \subset C^\infty(M \times \mathbf{S}^1)$ and $\mathcal{H} \subset H^1(M, \mathbf{R})$ such that $\psi \in \mathcal{A}$ and $c \in \mathcal{H}$ imply card $\mathcal{M}^c(L + \psi) = 1$.

The core of Theorem 20.13 is a), while b) is essentially a corollary of a) via standard arguments on the residuality of points of continuity of upper semicontinuous functions. An interesting problem posed by Theorem 20.13 is whether one can replace "residuality" by "full measure".

A result in a certain sense "dual" to Theorem 20.13 is the following:

Theorem 20.13'. *[Mñ, Th.C] Let L be a Lagrangian satisfying the same hypotheses as in Theorem 20.13. Then:*

a) For every $h \in H_1(M, \mathbf{R})$ there exists a residual subset $\mathcal{A}(h) \subset C^\infty(M \times \mathbf{S}^1)$ such that $\psi \in \mathcal{A}(h)$ implies $\operatorname{card} \mathcal{M}_h(L + \psi) = 1$.

b) There exist residual subsets $\mathcal{A} \subset C^\infty(M \times \mathbf{S}^1)$ and $\mathcal{H} \subset H_1(M, \mathbf{R})$ such that $\psi \in \mathcal{A}$ and $h \in \mathcal{H}$ imply $\operatorname{card} \mathcal{M}_h(L + \psi) = 1$.

Again, one may ask whether "residual" can be replaced by "full measure" in Theorem 20.13' or whether the set $\mathcal{H}$ in part b) of the theorem contains those h for which $\mathcal{M}_h(L + \psi)$ contains an ergodic measure. The last question can be reformulated as follows: are the minimal ergodic measures determined by their homology?

The proofs of Theorems 20.13 and 20.13' are contained in [Mñ], together with other results in the same spirit. We would also like to mention recent results obtained by the first author [Ma13] in the direction of extending to more degrees of freedom the variational construction of orbits sketched in the second part of §19 in the case of twist maps of the annulus. The main step is a definition of the appropriate variational principle and Peierls's barriers for Lagrangian systems, which generalize (at least partially) the corresponding concepts for twist maps. These generalizations are then applied to the construction of wandering orbits. However, the results obtained so far [Ma13] fall far short of what one would expect in more degrees of freedom.

References

[An1] S.Angenent: Monotone recurrence relations, their Birkhoff orbits, and their topological entropy, *Ergodic Th. Dynam. Sys.* (to appear)

[An2] S.Angenent: A remark on the topological entropy and invariant circles of an area preserving twist map, in *Twist mappings and their applications*, R.McGehee and K.R. Meyer editors, New York, Springer-Verlag, 1992.

[Au] S.Aubry: The twist map, the extended Frenkel-Kontorova model and the devil's staircase, *Physica* **7D** *(1983), 240-258.*

[Au-LeD] S.Aubry-P.Y.LeDaeron: The discrete Frenkel-Kontorova model and its extensions I: exact results for the ground states, *Physica* **8D** *(1983), 381-422.*

[B-M] J.Ball-V.Mizel: One dimensional variational problems whose minimizers do not satisfy the Euler-Lagrange equation. *Arch. Ration. Mech. Anal.* **90** *(1988), 325-388.*

[Ba] V.Bangert: Mather sets for twist maps and geodesics on tori, *Dynamics Reported* **1** *(1988), 1-45.*

[Bi1] G.D. Birkhoff: Surface transformations and their dynamical applications, *Acta Math.* **43** *(1922), 1-119.* Reprinted in *Collected Mathematical papers*, American Math. Soc., New York, 1950, Vol. **II**, 111-229.

[Bi2] G.D. Birkhoff: On the periodic motion of dynamical systems, *Acta Math.* **50** *(1927), 359-379.* Reprinted in *Collected Mathematical papers*, American Math. Soc., New York, 1950, Vol. **II**, 333-353.

[Bi3] G.D. Birkhoff: Sur quelques courbes fermées remarquables., *Bull. Soc. Math. de France* **60** *(1932), 1-26.* Reprinted in *Collected Mathematical papers*, American Math. Soc., New York, 1950, Vol. **II**, 418-443.

[Bl] S.Bullet: Invariant circles for the piece-wise linear standard map, *Comm. Math. Phys.* **107** *(1986), 241-262.*

[De] A.Denjoy: Sur les courbes définies par les équations differentielles à la surface du tore, *J. Math. Pures Appl.* **11** *(1932), 333-375.*

[F] G.Forni: *Construction of invariant measures and destruction of invariant curves for twist maps of the annulus*, Ph. D. Thesis, Princeton University, October 1993.

[Gr] J.M.Greene: A method for determining stochastic transition, *J. Math. Phys.* **20** *(1979), 1183-1201.*

[Hd] G.A. Hedlund: Geodesics on a two-dimensional Riemannian manifold with periodic coefficients, *Ann. of Math.* **33** *(1932), 719-739.*

[He] M.R. Herman: Sur les courbes invariantes par les difféomorphismes de l'anneau, Vol. I & II , *Asterisque* **103-104** *(1983) & **144** (1986).*

[La] V.F.Lazutkin: The existence of caustics for a billiard problem in a convex domain, *Math. USSR Izvestija* **7** *(1973), 185-214*. Translation of *Izvestija, Mathematical series*, Academy of Sciences of the USSR, **37**, 1973.

[L-L] A.J. Lichtenberg-M.A. Liebermann: *Regular and Chaotic Dynamics*, Springer-Verlag, New-York 1983 (Second Edition 1992)

[MK-P] R.S.MacKay-I.C.Percival: Converse KAM: Theory and Practice, *Comm. Math. Phys.* **98** *(1985), 469-512.*

[Mñ] R. Mañé: Properties and Problems of Minimizing Measures of Lagrangian Systems, preprint, 1993.

[Ma1] J.N. Mather: Existence of quasi-periodic orbits for twist homeomorphism of the annulus, *Topology* **21** *(1982), 457-467.*

[Ma2] J.N. Mather: Glancing billiards, *Ergod. Th. Dynam. Sys.* **2** *(1982), 397-403.*

[Ma3] J.N. Mather: letter to R.S. MacKay, February 1984.

[Ma4] J.N. Mather: Non-existence of invariant circles, *Ergod. Th. Dynam. Sys.* **4** *(1984), 301-309.*

[Ma5] J.N. Mather: Non-uniqueness of solutions of Percival's Euler-Lagrange equations, *Commun. Math. Phys.* **86***(1983), 465-473.*

[Ma6] J.N. Mather: More Denjoy invariant sets for area preserving diffeomorphisms, *Comment. Math. Helv.* **60** *(1985), 508-557.*

[Ma7] J.N. Mather: A criterion for the non existence of invariant circles, *Publ. Math. I.H.E.S.* **63** *(1986), 153-204.*

[Ma8] J.N. Mather: Modulus of continuity for Peierls's barrier , *Periodic Solutions of Hamiltonian systems and Related topics*, ed. P.H. Rabinowitz et al. NATO ASI Series C **209**. D. Reidel, Dordrecht (1987), 177-202.

[Ma9] J.N. Mather: Destruction of invariant circles, *Ergod. Th. Dynam. Sys.* **8** *(1988), 199-214.*

[Ma10] J.N. Mather: Minimal measures, *Comment. Math. Helv.* **64** *(1989), 375-394.*

[Ma11] J.N. Mather: Variational construction of orbits for twist diffeomorphisms, *J. Amer. Math. Soc.* **4** *(1991), no. 2, 203-267.*

[Ma12] J.N. Mather: Action minimizing invariant measures for positive definite Lagrangian systems, *Math. Z.* **207** *(1991), 169-207.*

[Ma13] J.N. Mather: Variational construction of orbits of twist diffeomorphisms II, to Bernard Malgrange on his 65th Birthday, preprint (to appear in the Proceedings of the Malgrange Fest).

[Mo1] J.Moser: *Stable and Random motions in Dynamical Systems*, Princeton Univ. Press, Princeton, 1973.

[Mo2] J.Moser: Monotone twist mappings and the calculus of variations, *Ergod. Th. Dynam. Sys.* **6** *(1986), 401-413.*

[P-deM] J. Palis-W.de Melo: *Geometric Theory of Dynamical Systems: An Introduction*, Springer-Verlag, New York-Heidelberg-Berlin, 1982.

[P-M] R.Perez-Marco: Solution complète au problème de Siegel de linéarization d'une application holomorphe au voisinage d'un point fixe (d'après J.-C. Yoccoz), *Séminaire Bourbaki, 44ième année,* **753** *(1991-92), 273-309.*

[Pe1] I.C. Percival: A variational principle for invariant tori of fixed frequency, *J. Phys. A: Math. and Gen.* **12** *(1979), No. 3, L.57.*

[Pe2] I.C. Percival: Variational principles for invariant tori and cantori, in *Symp. on Nonlinear Dynamics and Beam-Beam Interactions, (Edited by M. Month and J.C. Herrara), No. 57 (1980), 310-320.*

[Po] H.Poincaré: *Oeuvres*, Vol. I., Gauthier-Villars, Paris, 1928-1956.

[Rc] R.T.Rockafellar: *Convex Analysis*, Princeton Math. Ser., vol. 28, Princeton University Press, Princeton, 1970.

[Rs] H.Rüssmann: On the frequencies of quasi-periodic solutions of nearly integrable Hamiltonian systems, Preprint, Euler International Mathematical Institute, St. Petersburg, Dynamical Systems, 14-27 October 1991.

[S-Z] D.Salamon-E.Zehnder: KAM theory in configuration space, *Comm. Math. Helvetici* **64** *(1989), 84-132.*

[Sw] S.Schwartzman: Asymptotic cycles, *Ann. Math. II Ser.,* **66** *(1957), 270-284.*

[Yo] J.-C.Yoccoz: Conjugaison Différentiable des Difféomorphismes du Cercle dont le Nombre de Rotation Vérifie une Condition Diophantienne, *Ann. Scient. Éc. Norm. Sup.,* **17** *(1984), 333-359.*

S. ABENDA, ISAS/SISSA, Via Beirut 4, 34014 Trieste

A. BAZZANI, Dip. di Fisica, Via Irnerio 46, 40126 Bologna

E. CALICETI, Dip.to di Mat., Piazza di Porta S. Donato 5, 40127 Bologna

A. CHERUBINI, Dip. di Mat., Piazza di Porta S. Donato 5, 40127 Bologna

M. DEGLI ESPOSTI, Via Matteotti 13, 40069 Zola Predosa, Bologna

L. ERDOS, Princeton Univ., Dept. of Math., Fine Hall, Washington Rd.,
     Princeton, NJ 08544

G. FORNI, Via Putti 20, 40133 Bologna

J.-M. GHEZ, Centre de Physique Théorique, CNRS Luminy, Case 907, F-13288 Marseille

C. GIBERTI, Dip. di Mat. pura e appl., Via Campi 213/B, 41100 Modena

E. GUTKIN, Dept. of Math., Univ. of Southern California, Los Angeles, CA 90089-1113

E. HARRELL, School of Math., Georgia Inst. of Techn., Atlanta, GA 30332-0160

J. HERCZYNSKI, Inst. Mat. Stosow. Uniw. Warszawski, Banacha 2, Warszawa 59

P. LOCHAK, D.M.I., Ecole Normale Supérieure, 45 rue d'Ulm, 75005 Paris

M. LO SCHIAVO, Dip. di Metodi e modelli mat., Via A. Scarpa 10, 00161 Roma

M. MAIOLI, Dip. di Mat., Via Campi 213/B, 41100 Modena

S. MARMI, Dip. di Mat., Viale Morgagni 67/A, 50134 Firenze

F. NARDINI, Dip. di Mat., Piazza di Porta S. Donato 5, 40127 Bologna

A. PARMEGGIANI, Dip. di Mat., Piazza di Porta S. Donato 5, 40127 Bologna

T. PAUL, CEREMADE, Univ. Paris-Dauphine, Place du Marechal de Lattre de Tassigny,
     75775 Paris Cedex 16

M. STOLTZNER, Inst. f. Theoretische Physik der Univ. Wien, Boltzmanngasse 5,
     A-1090 Wien

E. TODESCO, Via Mascarella 77/V, 40126 Bologna

S. VAIENTI, Centre de Physique Théorique, Luminy, Case 907, F-13288 Marseille

M. VITTOT, Centre de Physique Théorique, CNRS, F-13288 Marseille

E. VRSCAY, Dept. of Appl. Math., Univ. of Waterloo, Waterloo, Ontario,
     Canada N2L 3G1

H. WU, SISSA, Via Beirut 4, 34014 Trieste

X. YE, Section of Mathematics, P.O.Box 586, 34100 Trieste

S. ZELDITCH, Dept. of Math., Johns Hopkins Univ., Baltimore, MD 21218

LIST OF C.I.M.E. SEMINARS   Publisher

1954 - 1. Analisi funzionale   C.I.M.E.
        2. Quadratura delle superficie e questioni connesse   "
        3. Equazioni differenziali non lineari   "

1955 - 4. Teorema di Riemann-Roch e questioni connesse   "
        5. Teoria dei numeri   "
        6. Topologia   "
        7. Teorie non linearizzate in elasticità, idrodinamica,aerodinamica   "
        8. Geometria proiettivo-differenziale   "

1956 - 9. Equazioni alle derivate parziali a caratteristiche reali   "
       10. Propagazione delle onde elettromagnetiche   "
       11. Teoria della funzioni di più variabili complesse e delle
           funzioni automorfe   "

1957 - 12. Geometria aritmetica e algebrica (2 vol.)   "
       13. Integrali singolari e questioni connesse   "
       14. Teoria della turbolenza (2 vol.)   "

1958 - 15. Vedute e problemi attuali in relatività generale   "
       16. Problemi di geometria differenziale in grande   "
       17. Il principio di minimo e le sue applicazioni alle equazioni
           funzionali   "

1959 - 18. Induzione e statistica   "
       19. Teoria algebrica dei meccanismi automatici (2 vol.)   "
       20. Gruppi, anelli di Lie e teoria della coomologia   "

1960 - 21. Sistemi dinamici e teoremi ergodici   "
       22. Forme differenziali e loro integrali   "

1961 - 23. Geometria del calcolo delle variazioni (2 vol.)   "
       24. Teoria delle distribuzioni   "
       25. Onde superficiali   "

1962 - 26. Topologia differenziale   "
       27. Autovalori e autosoluzioni   "
       28. Magnetofluidodinamica   "

```
1972 - 59. Non-linear mechanics                                              "
         60. Finite geometric structures and their applications              "
         61. Geometric measure theory and minimal surfaces                   "

1973 - 62. Complex analysis                                                  "
         63. New variational techniques in mathematical physics             "
         64. Spectral analysis                                               "

1974 - 65. Stability problems                                                "
         66. Singularities of analytic spaces                                "
         67. Eigenvalues of non linear problems                              "

1975 - 68. Theoretical computer sciences                                     "
         69. Model theory and applications                                   "
         70. Differential operators and manifolds                            "

1976 - 71. Statistical Mechanics                          Ed Liguori, Napoli
         72. Hyperbolicity                                                   "
         73. Differential topology                                          "

1977 - 74. Materials with memory                                             "
         75. Pseudodifferential operators with applications                  "
         76. Algebraic surfaces                                              "

1978 - 77. Stochastic differential equations                                 "
         78. Dynamical systems           Ed Liguori, Napoli and Birhäuser Verlag

1979 - 79. Recursion theory and computational complexity                     "
         80. Mathematics of biology                                          "

1980 - 81. Wave propagation                                                  "
         82. Harmonic analysis and group representations                     "
         83. Matroid theory and its applications                             "

1981 - 84. Kinetic Theories and the Boltzmann Equation  (LNM 1048) Springer-Verlag
         85. Algebraic Threefolds                        (LNM  947)          "
         86. Nonlinear Filtering and Stochastic Control  (LNM  972)          "

1982 - 87. Invariant Theory                              (LNM  996)          "
         88. Thermodynamics and Constitutive Equations   (LN Physics 228)    "
         89. Fluid Dynamics                              (LNM 1047)          "
```

1983 - 90. Complete Intersections	(LNM 1092)	Springer-Verlag
91. Bifurcation Theory and Applications	(LNM 1057)	"
92. Numerical Methods in Fluid Dynamics	(LNM 1127)	"
1984 - 93. Harmonic Mappings and Minimal Immersions	(LNM 1161)	"
94. Schrödinger Operators	(LNM 1159)	"
95. Buildings and the Geometry of Diagrams	(LNM 1181)	"
1985 - 96. Probability and Analysis	(LNM 1206)	"
97. Some Problems in Nonlinear Diffusion	(LNM 1224)	"
98. Theory of Moduli	(LNM 1337)	"
1986 - 99. Inverse Problems	(LNM 1225)	"
100. Mathematical Economics	(LNM 1330)	"
101. Combinatorial Optimization	(LNM 1403)	"
1987 - 102. Relativistic Fluid Dynamics	(LNM 1385)	"
103. Topics in Calculus of Variations	(LNM 1365)	"
1988 - 104. Logic and Computer Science	(LNM 1429)	"
105. Global Geometry and Mathematical Physics	(LNM 1451)	"
1989 - 106. Methods of nonconvex analysis	(LNM 1446)	"
107. Microlocal Analysis and Applications	(LNM 1495)	"
1990 - 108. Geoemtric Topology: Recent Developments	(LNM 1504)	"
109. H Control Theory	(LNM 1496)	"
110. Mathematical Modelling of Industrial Processes	(LNM 1521)	"
1991 - 111. Topological Methods for Ordinary Differential Equations	(LNM 1537)	"
112. Arithmetic Algebraic Geometry	(LNM 1553)	"
113. Transition to Chaos in Classical and Quantum Mechanics	(LNM 1588)	"
1992 - 114. Dirichlet Forms	(LNM 1563)	"
115. D-Modules, Representation Theory, and Quantum Groups	(LNM 1565)	"
116. Nonequilibrium Problems in Many-Particle Systems	(LNM 1551)	"

1993 –	117. Integrable Systems and Quantum Groups	to appear	Springer-Verlag
	118. Algebraic Cycles and Hodge Theories	to appear	"
	119. Phase Transitions and Hysteresis	(LNM 1584)	"
1994 –	120. Recent Mathematical Methods in Nonlinear Wave Propagation	to appear	"
	121. Dynamical Systems	to appear	"
	122. Transcendental Methods in Algebraic Geometry	to appear	"